DOCUMENTATION
BASICS

That Support
Good Manufacturing Practices
and Quality System Regulations

2nd Edition

Carol DeSain

Tamarack Associates, LLC
www.thetamarackgroup.net

Published by Tamarack Associates, LLC

Printed by Shel/Don Reproduction Centre Inc., Duluth, Mn

To purchase additional copies of *Documentation Basics, 2nd Edition*, order on-line at www.thetamarackgroup.net.

For questions contact us at books@thetamarackgroup.net.

Printed in the United States of America

About the Author

As a consultant since 1986, Carol DeSain works domestically and in Europe advising biologic, drug, diagnostic and device companies in the compliance management of their operations, start-up and renovation strategies, and quality system design and implementation (documentation, validation, investigations, specifications, change control). Carol is a biologist, having worked with bacterial, viral and animal cell systems enzymes in academic research, industrial development, start-up entrepreneurial companies and contract manufacturing. Over 30 years of experience has led Carol to an understanding of what individuals need to know to excel, professionally, in a regulated industry. She is well known for her practical approach to the regulatory challenges of complex operations, her interactive seminars, articles and books. Previous books include *Drug, Device and Diagnostic Manufacturing* (Interpharm), *Documentation Basics* (Advanstar); and, with Charmaine Sutton, *Documentation Practices* (Advanstar), *Validation for Medical Device and Diagnostic Manufacturing* (Interpharm), *Meeting GMP and ISO 9001 Expectations for Product Development* (Parexel), and *Risk Management Basics* (Advanstar).

*Ten years ago I dedicated the first edition of
Documentation Basics to Charmaine Vercimak,
Georgiann Keyport, and Diane Krizek.*

*Now, ten years later, I am pleased to confirm
that Charmaine, Georgiann, Diane and I remain
friends and colleagues, that we all still work in
the medical products industry, and that with age
we have all realized—more than ever—not to
believe beyond the documented evidence,
except in our support for one another.*

Contents

Section II—Documents

Exhibit E: Quality System Manual
Exhibit F: Standard Practice Document

Section III—Functional Area Documentation Basics

Introduction to the Second Edition

DOCUMENTATION BASICS was first published ten years ago. It was written because there was no other source of information available on the topic at that time. In addition, it was written to serve a practical purpose for start-up companies who were, in my opinion, wasting a lot of time and money designing part numbers, lot numbers and Standard Operating Procedures (SOPs).

Now, ten years later, with a wealth of innovative technologies available from medical developers and manufacturers, the industry (both start-up and established companies) continues to struggle with the paperwork. I have resisted rewriting this book expecting that:

- something better would come along;
- everyone would soon know all of this information instinctively;
- we, as an industry, would simply outgrow the paper; and finally,
- that *Documentation Practices*,[1] published in 1996 by Carol DeSain and Charmaine Sutton would replace the need for *Documentation Basics*.

Now, I have surrendered to the obvious because:

- the industry continues to purchase *Documentation Basics*;
- product development in device and diagnostic industries has emerged as a new area requiring the creation of controlled documents and records;
- new companies continue to emerge for device, biologic, dietary supplement, and genetic testing products—requiring new document systems;
- multinational companies and companies created from mergers and acquisitions need to integrate existing document systems and harmonize across sites;
- the established industry continues to need—in spite of the electronic enhancements— the ability to design good numbering systems and good documents from scratch.

In *Documentation Basics, Second Edition*, I have honored the content of the first edition, providing a minimum set of standards to consider when designing document or material identification numbering systems, or writing SOPs, specifications, quality manuals, logbooks, etc. In addition, I have included chapters on documentation basics from the perspective of specific functional areas of operations, such as Material Handling, Maintenance, Quality Control, Production, Validation and Quality Assurance—providing guidance on common resources, practices and procedures, and documents/records for each functional area.

Documentation Basics is a companion to *Documentation Practices*, as *Documentation Practices* provides greater detail about different types of documents and document/record management practices. *Documentation Basics* is a start and perhaps sufficient for the functional area manager, but Document Control personnel or managers from large companies will quickly find the need for *Documentation Practices*, as well.

Example documents are provided throughout the text as well as lists of suggested documents and records. Examples and lists are provided as "just a start"; they are neither complete nor sufficient.

Carol's Caveat

Too many times in the last ten years, after auditing a document system that is wholly inadequate or dysfunctional, I have had clients say, "but we did exactly what you said to do in your book." So, as a point of clarification:

- *a good document system does not guarantee a good quality system;*
- *a good document format, good document change control, etc., does not guarantee compliance with regulatory expectations;*
- *numerous review and approval signatures in no way guarantees accurate, appropriate or complete document content.*

Documents are written to link people with operational responsibilities. If the individuals do not know what their responsibilities are, if they have not been provided the resources and training required to fulfill those responsibilities, or they do not take their responsibilities and, therefore, their signatures seriously, then the documents and records of the corporation become a compliance liability and a barrier to operational efficiency and effectiveness.

Good products are not made reliably from unreliable parts. Safe, effective and reliable medical products cannot be made without reliable, accurate instructions. But accurate instructions can be compromised when they are not written for and usually by the individuals expected to use the documents routinely.

So...if you follow the instructions, guidelines and examples provided in this book, you will be able to meet some of the fundamental expectations of GMPs/QSRs. It is up to you, not the document system, to fulfill your fundamental responsibilities to the users of your products, the regulators and the product development data. Good documents will simply help you do that consistently as your company grows and changes.

Carol DeSain
www.boreal.org/tamarack

Section I

Identifiers

1

Identification Numbers

In GOOD MANUFACTURING PRACTICE/QUALITY SYSTEM REGULATION (GMP/QSR) environments, numbers are used to identify the resources required to develop, manufacture, test, and distribute products. Identification numbers are used to identify documents, equipment, materials, chemicals, component parts, packaging materials, labels, production intermediates or subassemblies, samples, cell lines, final products, etc. Numbers are used instead of names because they are convenient; numbers are convenient because in a regulated industry, the identity of items, products, equipment, documents, etc., must be recorded routinely during development, production, testing and distribution, and numbers are easier to record than complete item descriptions.

Although identification numbers must be accompanied by control numbers (see Chapter 2) and sometimes status indicators (see Chapter 3) to completely identify any specific item, identification numbers are the foundation of identity in a GMP/QSR environment. Identification numbers must distinguish between items based on quality differences or differences that could affect the quality or performance of the end product. When this occurs, identification numbers support quality assurance, as all identified items, documents, equipment, etc., can be linked to records that specify the items and provide a chronology of maintenance and change.

Material Identifiers

Material identifiers may be called part numbers, item numbers, material codes, stock numbers, etc. All identification numbers, however, MUST identify items uniquely. The definition of "unique" will differ from industry to industry and from company to company, depending on how the items are used. Material identification numbers are a basic tool of GMPs that support an item's controlled purchasing and inventory control. With these numbers the identity of any item can be linked to purchasing specifications; with these numbers the location and availability of identified items can be determined.

What should/could have material identification numbers?

Although there is no regulation stating that "there must be material identification numbers," it is very difficult to control the identity, purchase, receipt, and storage of materials without them. Material identification numbers should be assigned for all purchased, consumable materials used in the production, testing and distribution of a product, and for all materials used to maintain or support its production, testing or distribution. In addition, identified materials should include deionization resins and regeneration chemicals used for purified water systems, lubricants used for "clean room" equipment, gases used in production and the cleaning agents, etc., used in maintenance.

What criteria are used to distinguish one identification number from another?

Assign different material identification numbers to materials when they differ from one another. Materials are considered different from one another when the differences in their identity, quality, testing or storage characteristics are significant enough to adversely affect their use or performance, if not distinguished from one another. Because copper wire is different from a titanium wire, and sodium chloride is different from sodium cyanide, these materials would be assigned individual material identification numbers. There are some uses of sodium chloride, however, where the grade of salt is critical, e.g., when sodium chloride, ACS is acceptable but sodium chloride, USP is not. Similarly there are processes where a filter from Millipore is acceptable but a filter from Sartorius is not, or vice versa. The differences that distinguish materials from one another are product- and process-specific. These decisions, therefore, must be made by someone knowledgeable about the use of each material or item.

Do not assign different material identification numbers to the same material only because it is used in different areas of the facility, such as R&D, QC or Production. There are exceptions, however. When a material like sodium chloride, ACS is used for more than one purpose in the facility, (e.g., used for buffer preparation in the laboratory and used directly in the formulation of products), then two different material identification numbers should be assigned *because* the two salts will be tested differently, routinely, and are, therefore, different from one another. When different designated uses of the salt will require different testing commitments, assign different material identification numbers. The differences in quality testing of each sodium chloride, ACS are established in the associated specification documents (Chapter 5). With this differentiation, however, the laboratory sodium chloride could *never* be used in Production.

Do not assign different identification numbers to the same quality of material based on the cost or the quantity of an item. A numbering system based on accounting, financial and material inventory criteria will not serve the quality control requirements of GMP/QSR environments.

Naming Materials

Materials are named to distinguish them from one another. As suggested above, the characteristics that differentiate one material from another are unique in every company and the differences should be evident in the format of the item name. When naming an item, start with its most common characteristic. For example, "Filter, 0.2 micron, cartridge, 10", cellulose," or "Tubing, silicone, 3/8" OD." In spite of the sorting capability of electronic records, well-defined naming conventions will lead to useful reference lists.

Categories of Identification Numbers

Categorizing identification numbers facilitates (1) recognition of the numbers by users, and (2) segregation of materials into groups with common characteristics that can be used to provide consistency in the content of the associated specification documents (see Chapter 4). To categorize numbers, assign specific meaning to certain numbers or letters in the identification number configuration. Such categorization of numbers, "smart numbers," is useful as long as the categories serve only to extend the purpose of identification (see Chapter 4). For example, consider separate categories of material identification numbers for:

- general supplies
- purchased chemicals, media, buffers, gases
- purchased components, parts, filters, tubing, sterile disposables for production
- purchased packaging materials and blank label stock
- purchased labels and printed materials
- purchased software
- software developed in-house
- prepared/manufactured solutions, media, buffers, WFI, PW
- prepared/manufactured production intermediates
- prepared/manufactured components and/or subassemblies
- labels and printed material produced in-house
- production and testing cell lines

Identification of Processing Intermediates or Subassemblies

As manufacturing proceeds, there are often stopping points where a subassembly or a processing intermediate is created. When these intermediates or subassemblies can be evaluated and judged as acceptable or unacceptable before production proceeds, and they represent a logical stopping point in production, it is suggested that a material identification number be assigned. Intermediate or subassembly identification numbers are appropriate for any in-house preparation of materials, media, buffers, tubing sets, fermentation harvests, crude extracts, etc., that are ultimately used in or are part of the final product.

Traditionally these materials are identified by a distinct category of material identification numbers that is linked to specification documents providing QC testing and inspection requirements as well as traceability to production methods.

Material Identification Numbers for Supplies

Although controlled purchase of "general supplies" is not a regulatory requirement, it is usually a business requirement and it is convenient to include the item in the material identification numbering system to facilitate purchasing. When supplies are given material identification numbers, however, provide a mechanism that clearly identifies a supply from a controlled material, such as a distinct category of identification numbers. Supplies can then be moved directly into inventory without labeling or inspection. Establishing a unique series of material identification numbers for supplies is one way of designating this difference.

Who Assigns Numbers?

Material identification numbers should be assigned by the quality unit of the organization. New numbers are required when a new product is transferred into the facility or when there are changes in existing products, processes, vendors, testing commitments, etc. In the QSR environment, when products are developed by the same corporation that performs commercial manufacturing, the assignment of new numbers for new products can be done early in the product development to facilitate purchasing and manufacturing. When product development and commercial manufacturing are performed by separate organizations, material identification numbers are often assigned by QA.

Material Specifications

As will be discussed in Chapter 4, specifications are written to support all identified materials. Specification documents establish the identity, purchasing controls, material storage/handling controls, and quality controls (including sampling plans) for each material. Specifications will vary in their format, according to the type of material they are designed to control. The identity description of a chemical, e.g., formula weight, empirical weight, appearance, etc., will differ from the identity description of a component, e.g., dimensions, composition, etc.

Document Identification Numbers

All documents created for use in the development, manufacturing, testing and distribution of products should be controlled. A controlled document is identified, reviewed, and approved before it is used. Changes to controlled documents are also reviewed and approved before implementation; these changes are designated with document edition or version numbers (see Chapter 2). Document identification numbers, like material identification numbers, provide a convenient way to identify documents uniquely, and they can also be used to extend the identification of a document into useful categories of identity.

Document Identification Number Categories

Document identification number categories can be used to designate types of documents. A standard operating procedure, for example, is a different type of document than a protocol; a manufacturing record is different than a material specification. Document identification numbers, therefore, should be configured to differentiate between specifications, procedures, protocols, forms, manufacturing records, etc., as these documents differ in format, content, and review/approval requirements.

Document identification number categories can also be used to designate document ownership, providing an important attribute of control for the document and its associated process. All documents are written to establish processing requirements. Sometimes the process is a test method used in the laboratory, sometimes it is a procedure used in manufacturing, sometimes it is the accepted way to clean a tank, or to release product into the market. Unlike material identification numbers, a second level of document categorization based on the location of use and intended user within the corporation can serve an important control function. Document categorization based on intended use and intended user can provide an opportunity to assign responsibility for the process document and thereby improve document review/approval management. For example, consider the following document identification number scheme:

DSQ-M101	
where D	identifies this item as a document
where S	is the type of document (SOP)
where Q	is the department that owns the document (QC)
where M	is a subtype of QC documents, e.g., microbiology SOPs
where 101	is a unique document identifier for all the microbiological SOPs in the QC department

In this example, the QC microbiologist would be the owner of the procedure. This individual would be considered the method expert. All documents in the DSQ-Mxxx document category would require only three signatures: QC Microbiology, QC Supervisor/Manager, and Quality Assurance.[1]

In a large multinational, company with several divisions and sites requiring a harmonized approach to document control, additional identifiers for location are appropriate. For example:

CP-DSQ-M101	
where C	is the division of a multinational corporation
where P	is the specific location of the facility (Puerto Rico)
where D	identifies this item as a document
where S	is the type of document (SOP)
where Q	is the department that owns the document in Puerto Rico
where M	is a subtype of QC documents, e.g., Microbiology SOPs
where 101	is a unique document identifier for all the microbiological SOPs in the QC department of the PR facility

Document Files/Archives

Document identification numbers are supported by document-specific files and/or databases that contain the current version of the document, all former versions of the document, a history of document change, and a history of document distribution. These document archives and associated change records are regulatory requirements.

Equipment Identification Numbers

Equipment identification numbers provide a tool to link information about equipment usage to associated GMP/QSR control records on equipment installation, validation, usage, maintenance, and change. Equipment identification numbers, like document and material identification numbers, can benefit from categorization. Consider equipment identification information that extends the identity of equipment in a manner that supports its controlled use, e.g., differentiating:

- stand-alone equipment from equipment that is part of an equipment system;
- equipment that is qualified or validated for use from equipment that is not;
- equipment that is maintained by outside contractors from equipment maintained in-house.

Identify all equipment that comes in direct contact with the product during production. Identify all equipment that requires routine calibration, maintenance evaluations, or repairs to ensure its performance. Consider the use of equipment identification number configurations that support these differences, for example:

EM-PW-P01
where E indicates that this is an equipment identifier
where M is a location identifier (MT = Maintenance)
where PW is an equipment or equipment system identifier,
     PW = purified water system
where P is the type of equipment (P = pump)
where the digits are unique equipment identifiers

Equipment identification numbers are assigned for microliter pipettes, thermometers, incubators, balances, micrometers, gauges, chromatography resin columns, detectors, chart recorders, etc. Identification numbers are not usually assigned to common glassware, or to laboratory equipment that does not measure or provide for a specific performance (e.g., vortex mixers, drying cabinets, flasks and beakers).

In a large multinational, company with several divisions and sites requiring a harmonized approach to document control, additional identifiers for location are appropriate. For example:

BPEM-PW-P01

where B is the division of a multinational corporation
where P is the specific location of the facility (Puerto Rico)
where E indicates that this is an equipment identifier
where M is a location identifier (MT = Maintenance)
where PW is an equipment or equipment system identifier,
        PW = purified water system
where P is the type of equipment (P = pump)
where the digits are unique equipment identifiers

Equipment Files

All equipment identification numbers should be associated with or supported by an equipment specification or equipment identification card and/or an equipment history file. Equipment specifications or identification cards provide a detailed description of the equipment. They can contain or reference equipment usage logs, preventive maintenance, calibration, repair and change history for the equipment over time. These files are a resource for all information about equipment and equipment systems (see Chapter 9).

Sample Identifiers

Samples taken from materials, products, in-process manufacturing and testing events, utility systems, validation studies, etc., must be uniquely identified. Sample identifiers should be issued by QC. These numbers must be evident on the sample itself, and in the records directing the collection of the sample. Samples require a name or identity of the sample source (e.g., WFI sample, or DEAE eluate sample, or silicone gasket sample). In addition, sample identification must be linked to the sampling process. Sample identifiers, as a result, must be supported by a sampling document that provides the detailed description of the sample size, sampling location, sample time, and sample configuration.

Cell Lines: Master vs. Working Cell Bank Identifiers

Cell lines used in production and testing are materials that require identification. All cell lines, however, have two distinct levels of control that need separate identity: master cell banks and working cell banks. When the material identification numbering system cannot support this extended identity, it is appropriate to suffix the identity number with an M or W, as appropriate. The passage level of the cell line, however, is more appropriately designated in the control number, as discussed in the next chapter.

Using Identification Numbers in the GMP/QSR Environment

What is the difference between a production or laboratory area designated as meeting GMP/QSR regulations and one that is not? One obvious and visual difference is that in the GMP/QSR environment, everything is identified. In the GMP laboratory, for example:

- the area itself is identified as the QC Laboratory and is a restricted access area
- solutions are identified consistently with identification names/numbers, lot numbers or preparation dates, expiration dates, and storage conditions
- equipment is identified with equipment ID #s and calibration stickers
- laboratory notebooks are identified with ID #s
- samples are identified with identification numbers and dates
- refrigerators, freezers, incubators, test stations, storage cabinets, etc., are identified with their intended use

In a GMP production area, for example:

- production areas are identified to identify the current product and lot being manufactured
- all materials in the area are identified with identification number and lot numbers
- all materials are identified with status tags, indicating that the materials are available for use
- equipment is identified with equipment ID #s
- gas tanks are identified with identification #s, lot #s, and expiration dates
- spray bottles with alcohol are identified with identifiers, date of preparation, date of expiration
- control panels have calibration stickers associated with gauges and meters.

GMP/QSR environments are supported by information records that specify and confirm the consistent performance of all resources. Identification numbers link the resources to the records.

Reuse of Numbers

Identification numbers should not be reused. Once identification numbers have been used in records about the manufacture, testing and distribution of product, these records and the associated numbers remain open to review by investigators and the due-diligence proceedings of the legal system for many years. Do not reassign numbers, retire them if the documents, equipment or materials they describe are no longer required.

2

Control Numbers

COMPLETE IDENTIFICATION OF MATERIALS, product, and documents cannot be achieved with identification numbers alone. While material identification numbers, equipment identification numbers and document identification numbers designate identity, control numbers extend that identity based on time and/or location qualifiers. Control numbers locate the items in time and place; they designate when/where materials were received, equipment was used or documents were changed. Final product lot numbers distinguish identical products—that are manufactured at different times or during different cycles of production—from one another. Document version numbers distinguish different editions of a document—created, reviewed and approved at different times in the operational history of the company—from one another. Receiving codes distinguish different sets of materials—with different manufacturer's lot numbers or different shipment times—from one another.

The control numbers associated with identification numbers provide a history of material receipt, document change, product manufacture, or equipment use. Identification and control numbers are used together to identify most resources used in GMP/QSR manufacturing and testing.

Product Lot Numbers

The same products with the same identification codes are manufactured many times throughout a year. Product lot numbers serve to distinguish these products from one another. A different product lot number is assigned to each separate production event. The product lot number, as described in the Code of Federal Regulations (CFRs), must be used on all of the records that support the production, testing, stability, inspection, packaging, and distribution of that product.

> *"Control number means any distinctive symbols, such as a distinctive combination of letters or numbers, or both, from which the history of the manufacturing packaging, labeling, and distribution of a unit, lot, or batch of finished devices can be determined." 21 CFR 820.3*

> *"Lot or batch means one or more components or finished devices that consist of a single type, model, class, size, composition, or software version that are manufactured under essentially the same conditions and that are intended to have uniform characteristics and quality within specified limits." 21 CFR 820.3*

> *"Lot number, control number or batch number means any distinctive combination of letters, numbers or symbols, or any combination of them, from which the complete history of the manufacture, processing, packing, holding and distribution of a batch or lot of drug product or other material can be determined." 21 CFR 210.3*

> *"Batch means a specific quantity of drug or other material that is intended to have uniform character and quality, within specified limits, and is produced according to a single manufacturing order during the same cycle of manufacture." 21 CFR 210.3*

Product lot numbers should be configured to contain only location and time designations. For example:

J4020322
where J is designates the New Jersey facility
where 4 designates manufacturing line 4 at the NJ site
where 02 designates the year 2002
where 03 designates March, meaning that product manufacture occurred in 3/00
where 22 designates the 22nd product manufactured on that line in March

---

Product Serial Numbers

Some products are manufactured one at a time. Some products, although manufactured in lots or batches, are identified individually to support their tracking during use. These control numbers are usually called serial numbers. They are either synonymous for the product lot number or they extend the identification of the lot number into individual units.

Document Version Numbers

Over the operational history of a company, there will be several versions of every document. Complete identification of a document, therefore, requires two identifiers: the document identification number and an associated control number (e.g., revision, version or edition number). Version identifiers are assigned to a document when changes to an existing document are initiated and approved. All document version numbers are supported with records that describe the change and the rationale for that change.

Version numbers can be categorized to support the status of a document, as will be discussed in the next chapter. Document version identifiers, for example, A, B, C can be used for SOPs in the development laboratory before validation. Document version numbers of 01, 02, 03 designation can be used for commercial operations after validation.

Equipment Cycle or Run Numbers

When equipment is used routinely to process or test components or products, its usage should be designated with a run number or cycle number. These numbers serve as the control numbers for equipment operation. Autoclave cycle numbers, for example, are assigned for each cycle of material processed; HPLC run numbers are assigned each time the equipment system is used for analysis. These numbers are recorded in product processing and testing records.

Cycle and run number designations are supported with records that describe the conditions of processing and the materials processed. These records can be stored chronologically or by cycle number under the equipment identification number.

Material Receiving Codes

When an identified item is received into the facility, the event is recorded. The receiving clerk matches the item to its purchase order, locates the material identification number on the order, and records receipt of the item in a receiving logbook.

The receiving logbook assigns receiving codes for items chronologically. The receiving code is linked with information about the item (e.g., the identification number, an item description, amount received and its configuration, supplier, manufacturer, manufacturer's lot number, and purchase order number).

Receiving codes can be simple numbers or alpha-numerics. Receiving codes of AA001, AA002, AA003…AA999, AB001, AB002, and so on, are adequate for most operations. For example, when a bottle of sodium phosphate, monobasic, ACS grade, arrives and the purchase order indicates that the material identification number as DSQC203, the receiving clerk enters information from this shipment on the line in the log containing the next available receiving code (for example, AA303). The bottle is then labeled with both the material identification number and receiving code.

A separate receiving code is assigned to each separate manufacturer's lot and for each separate shipping event. When, for example, 20 bottles of sodium phosphate (DSQC203) arrive, 15 with a manufacturer's lot number of 4R89T and 5 with a manufacturer's lot number of 5Z89N; 15 bottles will receive one receiving code and 5 bottles will receive a different receiving code. If another shipment of 5 bottles of sodium phosphate (DSQC203) arrive the next day, and the manufacturer's lot number is 5Z89N, these 5 bottles are assigned a third receiving code.

To assure the control and documentation of identified materials when these materials are received into the facility, they should be routed through a central receiving department. This includes packages delivered to the receptionist by the postal service or overnight courier and packages that are received after hours.

Solution Preparation Control Numbers

Reagents and solutions prepared for general manufacturing use or laboratory use must be identified completely. Every solution must have both identification and control designations. Many small companies use the date-of-preparation as the control number. Although this might be sufficient in a small operation, a large facility where identical solutions can be prepared several times a day in more than one area will require an additional identifier. QC020425, for example, might be an assigned lot number for a solution prepared in QC in April 2004; it is the 25th solution prepared in that area. PR020425 would be a different solution prepared in Production in April of 2004.

The identification and control designations of every solution must be traceable to a document that provides evidence for its preparation and testing. In small companies, this might be a solution preparation logbook; in large companies, solution preparation events are often documented in specific preparation forms/records.

Media Preparation and Buffer Preparation Lot Numbers in Biologics

Media and buffer preparation in biologic production are critical manufacturing events that are documented on a preparation record, similar to a product manufacturing record. This document contains information on solution preparation, testing, storage conditions, and expiration dating. Media or buffer formulation records are used to assign both a media/buffer material identification number and a distinct preparation lot number. When these materials are used in fermentation or purification events, these identifiers will appear on the Bill of Materials.

Cell Line Passage Numbers

Cell lines are materials used to produce products or intermediates. Cell lines are identified with material identification numbers. As cells are transferred from T-flasks to roller bottles to fermenter tanks in preparation for an inoculation event, however, it is often necessary to designate and/or control passage levels. Any designation that differentiates one level of cell scale-up from another—such as passage levels (P1, P2, etc.)—is a control number for the cell line and supports the identification number to completely identify the cell.

3

Status Identifiers

STATUS IDENTIFIERS EXTEND THE IDENTITY of materials, products, documents, and equipment provided by identification and control numbers, to include the phase of processing associated with the use of the item. Status indicators are displayed prominently on materials, product or equipment while it is in use in the manufacturing or testing facility; these designations do not appear on distributed product. Status indicators are used to prevent the mix-up of similar materials in different phases of processing. Status designations are supported with records that describe the decision-making process associated with the status identification.

Material and Product Status Identifiers
Materials, once delivered, are processed through a cycle of quarantine, dispositioning, storage, dispensing, use and/or disposal. Products, once produced, are processed through a similar cycle of quarantine, dispositioning, storage, etc. Status indicators associated with materials and products include:
- holding
- quarantine
- release
- rejected

Status designations appear on storage cages in the warehouse and the identification tags or labels are attached to materials or product pallets. Status designations also appear within the inventory control system. As status changes, it is customary to place a release sticker/tag/label over but not completely covering the quarantine status identifier.

Document Status Identifiers

Every document requiring change is processed through a document lifecycle. Document changes are proposed, reviewed and edited, and when acceptable, approved as the next version of the document. When the document is reviewed, it is considered a draft; only when the document is approved for use is it considered a final document. Status indicators of "draft" must be evident on documents. Some companies prepare draft documents on one colored paper and originals on another color, leaving true copies for white paper; some companies will watermark documents.

Area Status Identifiers

Environmentally controlled areas and production/ testing suites are processed through cycles of use and preparation that must be evident to all users. Phases of processing that should be identified with status indicators include:

- available for use
- in use
- available for cleaning
- locked-out

Equipment Status Identifiers

Equipment can exist in many stages of use or repair without its status being obvious to an observer. Equipment status tags, as a result, are a good practice. Consider equipment tags for:

- available for use
- in use
- available for cleaning
- locked-out

Warehouse Status Identifiers

Materials and products stored in the warehouse have status indicators aside from the designations mentioned previously for holding, quarantined, released, and rejected. These status identifiers are associated with the inventory control and storage characteristics of the items. Consider status identifiers for:

- opened boxes/pallets as "partials"
- sampled boxes or containers as "sampled"

Processing Identifiers

During product production, there are many in-process steps where intermediates are stored or sampled. Complete identification of tanks or bins containing these materials, or complete identification of samples taken from these tanks or bins requires an extension of the identification and control numbers of the batch or lot. These extensions, usually designated in batch-processing records, provide a status for the in-process materials that is essential for their identification. For example, when a sample is taken at step 24 in the production of Product XYZ; Lot AD10327, the sample can be identified as XYZ; AD10327 - 24.

Use of Material and Product Identifiers

Identification numbers, control numbers and status identifiers *must* be used to support GMP/QS regulations. Developing good visual status indicators is essential when they are used to prevent the mix-up of similar-looking materials in different stages of processing. Labeling of materials, products, equipment, documents, and processing areas must be performed rigorously and consistently.

Materials/Products

Each item received is identified with its material identification number, receiving code, storage conditions, and a quarantine tag. The word "quarantine" is customarily located at the top of the label and is usually orange in color. As the status of the material changes throughout its lifecycle the labels change. This is discussed in Chapter 7.

Processing Areas

Processing areas require identification. These can be handwritten signs or display panels that appear at the entrances to the areas designating the purpose of the area and any activities that are currently performed there. As with materials, color coding can be a useful attribute of this signage.

Equipment

Equipment identifiers must be located where they are easy to see when the equipment is being used. They should be permanent identifiers since these locations are often subject to adverse moisture and temperature conditions.

Section II

Documents

4

Specifications

A SPECIFICATION DOCUMENT is the primary source of information for an item. It is a written, detailed description of the item and the controls applied to its purchase, production, storage or testing while in the hands of the company. As such, when an item is assigned a material identification number, this assignment is supported by a specification document. When an identified item is released for use in processing or testing (see Chapter 7) the item has been evaluated and it has been determined that it meets the requirements of the specification document. As a result, sodium chloride, ACS purchased from J.T. Baker becomes corporate part #344. Corporate part #344 is sodium chloride, ACS that has been purchased, inspected, stored, and tested according to corporate specification 344.

Specification documents provide the descriptive information that distinguishes an item from other similar items in the department, the facility or the company. Drawings, used primarily in the device industry, are also considered specifications *if* they contain the appropriate identification and control information.

This chapter provides an abbreviated description of specification documents. Greater detail is provided in *Documentation Practices*.[1]

Specification Document Categories

As presented in Chapter 1, specification documents can be appropriately segregated into categories that facilitate material identification. Purchased chemical specification documents, for example, would require entries for formula weight, empirical weight, appearance, and approved vendors. Prepared solution specification documents would require actual recipes or references to recipes, and solution characteristics such as optical density, viscosity, color, clarity, etc. When specification documents are designed to contain item-specific descriptors or identifiers, the following categories should be considered:

- general supplies
- purchased chemicals, media, buffers, gases
- purchased components, parts, filters, tubing, sterile disposables for production
- purchased packaging materials and blank label stock
- purchased labels and printed materials
- purchased software
- software written in-house
- prepared solutions, media, buffers, WFI, PW
- prepared production intermediates
- prepared component subassemblies
- labels and printed material produced in-house
- production and testing cell lines.

Specification Document Format

All specification documents should provide the opportunity to establish four general types of control for an item:

- item identification characteristics, features, or grade
- purchasing or production controls
- material controls
- quality controls

Any of this item control information could be used to distinguish similar items from one another. As suggested, specification documents can be written from templates; the design of these templates could be different for each category of specification (see Exhibit A). Consider the following guidance when designing specification documents.

Header/Format

- the name of the company
- a title or narrative description of the item
- the material identification number of the item
- the edition number of the specification (not the form)
- pagination (page 1 of 3, for example)
- when appropriate, form ID # and form version #
- specification approval signatures

Item Description Information Section

- Chemical specification forms can contain a physical appearance description of the chemical, chemical grade, empirical formula, and formula weight.
- Component and subassembly specification forms can contain sections on material composition, size, dimensions, color, and may reference an attached drawing.
- Cell-line specification forms can indicate whether the cell line is suspension- or anchorage-dependent and detail cell-line history, vectors, markers, passage level limits, isoenzyme analysis, etc.
- Printed materials or labeling specification forms can contain a color chart (when appropriate) and an actual approved master copy of the printed item to use for comparison during release work.
- Solution specification forms can contain information such as formulation instructions, final solution appearance, pH, viscosity, and specific gravity.

(Note: This information will vary according to the category of the item.)

Purchasing/Production Control Section

- For *purchased items*, provide a list of approved vendors for that item with catalog numbers.
- For *items prepared/manufactured in-house*, provide a reference to the manufacturing process for that item (SOP # or Batch Record ID #).

Material Control Section

- an expiration date for the item
- sample size (when testing is required)
- file sample size (when required for critical items)
- storage conditions
- handling and safety precautions
- requirements for shipping paperwork (e.g., Certificates of Analysis)
- inspection requirements for receipt of material

Quality Control Section

- specific requirements for Certificate of Analysis review
- actual testing required; list the
    - test parameter
    - test method ID
    - number of items tested or sample configuration
    - acceptance criteria

Specification Approval Signatures

Specification documents must be approved. Approval is granted by signing the document. Who should sign specification documents? It depends. If you are able to categorize specifications, as suggested above and in Chapter 1, then this categorization can limit the number of signatures required to those functional areas responsible for performing the work associated with the item. Consider the following suggestions:

Specifications	Approval Signatures
Purchased Material	QA + Purchasing + QC
Materials produced in-house	QA + Production + QC
Production intermediates/subassemblies	QA + Production + QC
Final Product	QA + QC + Regulatory
Labels and Labeling	QA + QC + Regulatory

Specification Content

Specification documents are not easy to write because specifications are the baseline against which changes and deviations are measured or observed over time. Although specifications should have been developed for the product and its processing through the Product Development process, many specifications are in fact developed on the manufacturing floor or in the QC laboratory for support materials and processes. For information on how to establish specification testing commitments, consult *Documentation Practices*.[1]

All identified materials should be described in a specification document and its purchasing or production control established. Some identified materials might require additional controls to ensure quality over shelf-life. Controlled storage conditions and QC testing should be used when appropriate to help ensure consistent performance of resources and products.

Specification Files/Archives, Distribution, and Usage

Approved copies of current specification documents must be readily available for use in the areas of the facility where items are produced, purchased, received, sampled, or tested. Only one version of a specification document should be available for use at any time.

Specification document files—like any document file—should contain the current, approved, original copy of the specification document, a document history log or file detailing the revision history and rationale for document change, a document distribution log, and copies of all former revisions of the specification document. These files must be secure with limited access.

Tofte Medical, Inc.

Page 1 of 1

Specifications for Chemicals

Form 79; 01

Name: Sodium Chloride, ACS

Part Number: 2176
Spec. Revision #: 01

Usage Requirements: Laboratory chemical for buffer preparation

Quality and Identity Characteristics

Appearance: white cube-like crystals

Empirical Formula: $NaCl$
Formula Weight: 58.44

Purchasing Control:

Purchasing Requirements: ACS grade; available in 500 gram bottles;
Certificate of Analysis available on bottle

Approved Vendors	Catalog #
Mallinckrodt	7581-500
J.T. Baker Chemical	3624-01

Material Control:

Receive, label and quarantine material according to SOP 765.
Expiration Date: 5 years   Storage Conditions: 20-25C
Handling Precautions: Avoid high humidity conditions

Quality Control:

Sample material according to SOP 098.
Sample Size: 1 gram   File Sample Size: NA

Testing	Method/SOP	Acceptance Criteria
Sodium ID	SOP 425	positive for sodium
Chloride ID	SOP 425	positive for chloride

Specification Approval Signatures:

QC_____ Date _____ Purchasing _____ Date_____
QA_____ Date _____

Exhibit A

5

Standard Operating Procedures/Forms

STANDARD OPERATING PROCEDURES (SOPs) are documents that describe how to perform routine tasks in the GMP/QSR environments of product development, purchasing, production, testing, maintenance, material handling, quality assurance, and distribution. They contain step-by-step instructions that technicians and operators consult daily to perform their work reliably and consistently.

Data collection forms are documents a technician or operator completes while performing the routine tasks directed by the SOPs. Forms provide fill-in-the-blank spaces for the collection of raw data entries. Logbook entries, data printouts, and reports also provide evidence that the work proceeded as directed.

SOP Format

A basic format for SOPs is provided in Exhibit B and below; a more detailed description of SOP format and content is provided in *Documentation Practices.*[1]

Title. The SOP title should be brief and direct, describing each procedure in a way that identifies the process and the object of the process. Consider titles such as "Operation of (process) KMS Boilers (object)"; Testing of (process) Buffer Solutions (object) for pH"; "Calibration of (process) WWT-60 Spectrophotometers (object)".

1.0 Purpose/Scope. The purpose in an SOP format often restates a well-written SOP title, but it can also be used to expand upon or qualify the purpose of the procedure. Scope describes what the SOP does and does not apply to. For example, if an SOP describes the calibration of a spectrophotometer, the procedure might include only spectrophotometers in the

QC laboratory, only spectrophotometers used in GMP operations, only spectrophotometers used for potency assays, or only double-beam spectrophotometers. To declare the scope of an SOP, therefore, consider what the procedure applies to, the individuals it might apply to, and when it is applied.

2.0 Responsibility in an SOP format simply declares who is responsible for performing the operations cited. It might cite a department or mandate specific training requirements for individuals within a department.

3.0 Established Process. Procedures should be written in simple, direct, step-by-step language that explains how to perform the tasks.[1] They should be written in a language that the user of the SOP will understand. When possible, use diagrams and visuals to support the narrative.

Each page of an SOP document should contain the company name, the title of the SOP, SOP identifier, version number, and pagination (for example, page 1 of 12). The company name and some declaration of confidentiality can also appear on each page. As long as each page contains the SOP identifier and its version number, only one page—usually the first—needs to contain dated document approval signatures.

Who Writes SOPs?

SOPs support routine GMP/QSR-regulated operations. They tell how the task will be performed (procedure), who will perform it (responsibility), why it will be performed (purpose), and what, if any, limits of use apply (scope). SOPs should be written by or with the individuals who perform the operations.

SOP Language and Detail

Do not write SOPs for the FDA. Write SOPs for the technicians and operators who will use them on a daily basis. Use clear and direct language. Use active verbs for procedural directives: "add this," "pour that," "observe…". When citing the use of an item in a procedure, name the item and cite its identifier (e.g., part number). Company slang terms for equipment, departments, or procedures can be used—as long as they are explained somewhere in the text to ensure clarity for an outside reviewer, such as FDA.

SOPs must be specific enough to be clear and accurate, yet flexible enough to be useful. Unnecessary detail can render a procedure useless before it is even approved. A directive can be specific, yet truly uninformative. Here is an example:

"Place the test tubes in the water bath in room G28 for 30 minutes.

Spin the tubes in the IEC centrifuge at setting #5 for 20 minutes.

Pour off the supernatant and dialyze it for 24 hours in cold WFI."

Although technicians might be able to perform those tasks as directed, any change—even a minor change—will require a change in the SOP. Furthermore, important details that support process quality control are missing from these instructions.

A directive such as "Place the test tubes in the water bath in room G28 for 30 minutes" should be changed to "Place the tubes in a 35-39° C water bath for 30 minutes (+/- 5 minutes)."

The action directed by this statement is to heat the tubes for an established amount of time at an established temperature. This is all that needs to be directed. It must, however, be directed specifically.

"Spin the tubes in the IEC centrifuge at setting #5 for 20 minutes" is also deficient as a directive. The intent of this step is to expose the tubes to an established amount of centrifugal force for an established period of time. Although that is accomplished at setting #5 in this IEC unit, there is no flexibility for the use of another, equally adequate centrifuge. Instead, state the controls for process parameters and add an example that directs the technician in the most likely method for fulfilling those processing expectations (e.g., "Spin the tubes at 8,000-10,000 xg for 20-25 minutes at 2-8° C [example: #5 setting on IEC centrifuge A267])."

"Pour off the supernatant and dialyze it for 24 hours in cold WFI" is specific, but not in a way that supports process control. Dialysis events should indicate the identification of the approved dialysis membranes, as well as a way to know when the dialysis is complete. Although temperature parameters must be established for dialysis events to ensure product stability, time parameters are related to the rate of dialysis and, therefore, to water volume and exposure time. Consequently, it is more appropriate to indicate the dialysis end point and then cite an example of water volume and time requirements.

SOP Review and Approval

There should be at least two, preferably three, signatures on an SOP. The person who owns the SOP should sign it first; this person is the one most knowledgeable about the work directed in the document and can attest to its accuracy. A second individual, usually in the same functional area as the owner, can approve the document; this individual should be responsible for assuring that the resources are available to do the work as directed. Finally, QA must review and sign all SOPs, providing the link to regulatory commitments, validation commitments and other functional area commitments that might be affected or compromised by the new SOP. Although many companies require additional signature blocks on a procedure, it is advisable to minimize the number of signatures and to establish the specific responsibilities of each document approver.

SOP Topics

Required topics for SOPs appear in the Code of Federal Regulations, Title 21, GMPs for finished pharmaceuticals (Parts 210, 211), GMPs for blood and blood components (Part 606), and QSRs for device and diagnostic products (part 820). These citations provide a *minimum* list of procedures. In addition, write an SOP that describes the initiation, approval, distribution, use, and change control of SOPs and data collection forms (see Exhibit B).

Data Collection Forms

As stated above, data collection forms are documents completed by a technician while performing routine tasks. The forms follow the instructions of the SOP while providing blank spaces for data entries.

In traditional research laboratories, laboratory notebooks are used to record data. The entries are informal—seldom are reagents identified completely, rarely is equipment identified, and signatures are missing. This is unacceptable for GMP/QSR manufacturing. Forms are created, therefore, to ensure that all information that must be documented while performing a procedure is, in fact, recorded.

Forms should be designed to facilitate the work they are used to record. The entry blanks should appear in the order in which the work is routinely performed and should require the technician to do as little writing as possible. Short, fill-in-the-blank entries are appropriate for receiving code entries, lot numbers, time, temperature, equipment identification numbers, room numbers where work is performed, calibration values, raw data values, and calculations.

Each page of a data collection form should contain the name of the company, the identity of the form, the form edition number, an appropriate SOP reference, pagination, and blank spaces for two signatures/dates (one for the technician who performs the work and one for a verification signature that indicates the form has been reviewed for accuracy, completeness, and compliance). See Exhibit C.

Tofte Medical, Inc.

Procedure AD-001, draft A

SOP 25; 01

Procedures: Creation, Review, The Approval, and Change of SOPs

1.0 Purpose/Scope

This procedure describes how to create, review, approve, and change a procedure. It describes the format and content requirements for all procedures, as well as document identification number assignments and title requirements. This procedure applies to all departments at TMI. ·

2.0 Responsibility/Training

This procedure is used by TMI employees and contractors trained to create, change, review, or approve procedures (see Training Modules 24-1, 24-2, 24-3, 24-4). Individuals expected to be trained in this procedure include all Departmental Management, designated document owners, and all QA and Document Control personnel.

3.0 Procedures

3.1 Introduction/Preliminary Operations/Requirements

Procedures are directive documents, they describe "how to do something." They should describe the performance of a routine task that requires documentation and data collection. The creation, review, approval, and change is a controlled, systematic process.

3.2 Flow of Work

3.2.1 New Procedures

When someone decides that a new procedure must be written or an existing procedure must be revised, they must:

- Identify the process/document owner for the procedure. Who will be responsible for the proper performance of this procedure, routinely? Who is the expert? Who is best qualified to be responsible for the accuracy of the procedure? Inform that individual of the need to create or change the procedure. Departmental management is responsible for assigning process owners.

- Complete a Document Initiation/Change Request Form 33, and submit it to Document Control. Document Control will assign a Document ID # (for new documents), a temporary revision level (1a, 2a, 3a, etc.), and a DCO# for tracking the document during its processing.
  Note: Information and document titles on the tracking forms are temporary; changes may occur before the document is approved.

Exhibit B

Tofte Medical, Inc.   Page 2 of 6

Procedure AD-001, draft A   SOP 25; 01

Procedures: Creation, Review, The Approval, and Change of SOPs

- Draft the document/procedure; label all drafts of that document with the assigned identifiers. This document can be drafted by anyone given the authority to do so by the process/document owner, but it should be reviewed by the process owner and departmental management, informally, i.e., not a documented review, before it is submitted to Document Control for official entry into the word processing system.

- Submit the draft procedure with the Document Initiation/Change Form 34 to Document Control. When information on the original initiation form changes, Document Control must make the required adjustments to title, number assignment, reviewers, etc.

- Document Control enters the procedure in their word processing system, their document log, and initiates document review by completing a Document Routing Form 35, and routing the document to: Quality Assurance, Department Management, Document Owner.

- The process owner is responsible for resolving all review comments and submitting a final draft of the document to Document Control for approval. The final documents are routed with an Approval Form 36.

- The document is signed and returned to Document Control. Consult AD-222 for document distribution requirements.

3.2.2 Titles

Procedure titles should be as descriptive as possible and present the most common aspect of the procedure first. For example, a procedure on the operation of the labeling machine should be titled "Operation of the Labeling Machine." Other appropriate titles include:

Inspection of…
Handling of…
Calibration of…
Cleaning and Assembly of…
Testing of…

When the object of the procedure can be specified exactly, do this in the title. If, for example, the procedure is titled "Cleaning of the Filling Machine," but the procedure is specific for one type of filling machine, retitle the procedure as "Cleaning of the Cozzoli XCV Filling Machine."

Exhibit B

Tofte Medical, Inc.   Page 3 of 6

Procedure AD-001, draft A   SOP 25; 01

Procedures: Creation, Review, The Approval, and Change of SOPs

3.2.3 Document Identification

All procedures will be identified by a unique 5-unit, sequentially assigned alphanumeric, i.e. QA-175. The department most responsible for the performance of the procedure will be identified in the numbering system. Document Control records the preliminary title of the document in the document identification number assignment log next to the number assigned.

3.2.4 Department Designations

AC   Accounting
AD   These are the procedures that apply to all departments: e.g.,
       AD-100 are the document control procedures;
       AD-200 are human resources/employee training procedures; and
       AD-300 are management procedures.
EM   Engineering/Maintenance
IS   Information Systems
MF   Manufacturing
MM   Materials Management
PD   Product Development
PR   Purchasing
QA   Quality Assurance
QC   Quality Control
RA   Regulatory
SM   Sales/Marketing

3.2.5 Procedure Creation, Review, and Approval

Procedures should be written by or in collaboration with the individuals who perform the work described. Procedures can be reviewed by more individuals than sign the final document. Appropriate reviewers are determined by the process/document owner and QA; reviewers other than the individuals who will approve the procedure are designated on the document initiation/change form. All procedures, are signed by three individuals, two individuals knowledgeable about the use of and/or performance of the procedure and a Quality Assurance signature.

The first signature is the process/document owner. This individual is responsible for:

- determining when/if the document requires change or when/if someone else's concerns about the document warrant changing it
- facilitating the change request and change review process; this work can be delegated but it remain the responsibility of the process owner

Exhibit B

Tofte Medical, Inc.   Page 4 of 6

Procedure AD-001, draft A   SOP 25; 01

Procedures: Creation, Review, The Approval, and Change of SOPs

- the accuracy of the information in the document. For example, that:
  - the document accurately describes the work as it will be done
  - part numbers, document ID numbers, equipment numbers, etc., are accurate
  - spelling and technical terminology are accurate
  - document format complies with corporate. guidance
- ensuring that the document is logical and written in a manner that is easily understood by those who perform the work
- consulting with other individuals or departments that might be impacted by the change or have an interest in the process being changed

The second signature is another individual knowledgeable about the work described in the document, usually departmental management in the same area as the document owner. The second signature is responsible for:

- verifying the accuracy of the information in the document
- verifying that the document is logical and written in a manner that is easily understood
- verifying that change is warranted
- verifying that other individuals or departments that might be impacted have been consulted
- ensuring that current industry practice has been followed and that, when available, industry standards or harmonized standards have been used
- ensuring that resources are available to complete the work as described (i.e., equipment, trained employees, processing areas)
- ensuring that review by other interested parties has occurred
- assuring that there are no conflicts with regulatory and validation commitments

The QA signature is responsible for:

- verifying that the document type and format meet the requirements of corporate document control procedures;
- verifying that regulatory and validation issues have been considered/resolved;
- verifying that changes do not compromise any other promises or commitments made to regulators;
- verifying that review by other interested parties has occurred; and
- ensuring that regulatory or harmonized standards, appropriate to the work, have been consulted and followed.

Exhibit B

Tofte Medical, Inc.   Page 5 of 6

Procedure AD-001, draft A   SOP 25; 01

Procedures: Creation, Review, The Approval, and Change of SOPs

3.2.6 Distribution, Filing, and Archiving

This procedure requires the creation and maintenance of active document files, containing the current approved version of the document, a change history log and a document distribution log; consult AD 233.

3.2.7 Procedure Revision, and Change Control

When a procedure requires change, the process owner is notified. A current copy of the procedure is marked up and it is submitted to Document Control with a Document Initiation/Change Form 33. The review of change and the subsequent approval process is identical to that presented in this procedure.

3.2.8 Procedure Format and Content Guidelines

Document Format: There should be a running header and footer on every page that contains the name of the company, pagination, the procedure identifier, and version identifier. In addition, the first page of every procedure should contain an document approval signature block.

There are 3 required sections of a procedure; they should be numbered as follows. The sections are:

1.0 Purpose/Applicability: This section restates the title and extends the description to include a detailed account of what the procedure will contain and/or what it will achieve. Also describe what the procedure applies to and/or what it does not apply to. For example, if a procedure describes the calibration of a sensor, it may apply only to the sensors used for a certain purpose or only to sensors used in the QA department. If a procedure describes an inspection procedure, it may apply only to a few products but not to all products. This selectivity should be described under scope.

2.0 Responsibility/Training This section describes who is responsible for the performance of the procedure. This could be an individual, a department or a group of specifically trained individuals. If specific training is required by the individuals authorized to perform the work, cite that training here.

Exhibit B

Tofte Medical, Inc.

Procedure AD-001, draft A

Procedures: Creation, Review, The Approval, and Change of SOPs

3.0 Procedure

3.1 Introduction/Preliminary Operations/Requirements *What needs to be known or done before work begins? Describe any calibrations of equipment, preparation of reagents, equipment suitability tests, sample preparation, etc. Format the information in this section, as applicable. This is an appropriate place to list any material, environmental or document requirements to assure before the work begins.*

> *Note: When referring to actual part numbers of materials that fulfill these characteristics, indicate if these part numbers are examples or requirements.*
>
> *"Sodium Chloride, ACS (for example PN 10-007)" indicates that I can use 10-007 or any other part number for Sodium Chloride which meets the ACS designation.*
>
> *Similarly designate equipment and equipment characteristics that must be met, (e.g., "UV-Visible Spectrophotometer with a 1cm path length (use only validated equipment #AV-78-C-01)").*

Environmental requirements can include clean room, cold room, incubator, and restricted access conditions.

List any associated procedures and data collection documents, such as forms and batch processing records used to fulfill the expectations of this procedure.

3.2 Flow of Work *This should be a step-by-step, chronological account of what to do. Sentences should be directive statements like "Mix this, Add that, Test those, Measure these." The procedure establishes the steps in a process and describes how to control the process to ensure its consistent performance.*

Describe the processing steps. What is the first step? What is done next? What materials and equipment are used, specifically (cite identification numbers, when appropriate)? What are the processing endpoints for each step? How do you know when processing is complete? Can the endpoint be measured?

3.3 Data Management and Notification Requirements *When the work has been completed, as directed in the procedure, what must be done with the data? Where is it recorded, how is it calculated, and who is notified of the result, and how?*

Exhibit B

Tofte Medical, Inc.

FM 905; A

Document Change Control Summary Form

○   **Document #** _____   **Proposed Revision** _____   **DCR#** _____ - __

(to be assigned by Documentation for new documents)

Title _____

Document type: (   ) SOP; (   ) Specification; (   ) Batch Record; (   ) Protocol;
                     (   ) Form; (   ) Other _____

(   ) New document:
                 (   ) Transferred from development (suffix = O)
                 (   ) Transferred directly from client (suffix = O)
                 (   ) Client = _____
                 (   ) Other = _____
(   ) Change to existing document; attach FM 905.

○   (   ) Change initiated from Emergency Change Order (suffix = E)
(   ) Periodic review of document requested (suffix = D)
(   ) Document a candidate for retirement/obsolescence (suffix = X)

Document Usage

         (   ) TMI Products

         (   ) Client = _____

             Client reference document used = ID# _____   Version _____

Document Owner or first signature on existing document = _____

○   Form completed by _____   Date _____

Exhibit C

Tofte Medical, Inc.

FM 905; A

Document Change Control Summary Form

Page 2 of 2

Form 33; 01

○   **Summary of Proposed Changes:**

    (   ) No changes; new document

    (   ) Clerical, clarification or administrative changes. No review required; submit to Documentation for approval signatures.

    (   ) Other _____

_____

Rationale for Changes: (Note: If change is an action implemented as a result of an investigation or a regulatory audit, provide reference to these activities.)

    (   ) Action resulting from Failure Investigation_____

    (   ) Regulatory commitment made; see_____

○   (   ) Action resulting from Product Development; see _____

    (   ) Other _____

Evaluation of Impact of Proposed Changes:

    (   ) No additional evaluation/training required

    (   ) Validation/Regulatory submission required before implementation

    (   ) Evaluations/Training required before implementation

    (   ) Evaluations/Training required after implementation

    (   ) Other _____

Explanation or references_____

○   _____

Form completed by _____ Date _____

Exhibit C

6

Logbooks

A LOG IS A CHRONOLOGICAL RECORD of activities generally associated with processing area usage, equipment usage or the flow of materials, product, information or samples through processing areas. Logbooks are used in any situation where deviations from the chronology of events could adversely impact product quality.

Logbooks provide access to information about the status of an area or an equipment system when status is not otherwise apparent. Logbooks can also provide a chronological listing of activities, such as material receipt, correspondence with FDA, document change, etc. When there are multiple uses or multiple users of an area or an equipment system, logbooks provide a record of the processing events that have occurred chronologically in an area or within a system, so that at any given moment one can determine the status of an area or a system by looking at the book.

Logbooks as Controlled Documents and Records

Logbooks are controlled documents used in every functional area of the corporation and should be issued and retrieved in a controlled manner. Logbooks can be assigned to equipment, locations, functional areas, or individuals; there should be an owner designated for each book. Logbooks should be bound in a manner that makes it difficult to lose or reorder pages; when appropriate, use waterproof ink and water-resistant papers. Logbooks can also be created by binding a set of forms together and assigning run #, cycle # or serial #s to the forms before they are bound into the logbook format. This approach provides for more specific data entry expectations in the logbook.

Minimum logbook entries include date, time, technician, and event. Other appropriate entries depend on its usage. A boiler logbook, for example, might list an established number of routine tasks, such as start-up, top blow-down, bottom blow-down, shutdown, and sampling. The list permits the mechanic to simply check off, sign, and date the event. Similarly, when a technician monitors equipment operating parameters such as line pressures, flow rates, temperature, and conductivity, the equipment logbook is a convenient place to record data. Logbooks are also appropriate for analytical equipment used for several different purposes by multiple individuals or departments. For example, equipment calibration, cleaning, column changeover, reference profiles, and sample runs are examples of appropriate logbook entries for an HPLC unit.

When logbooks are completed, they become records. They should be returned to the record archives for secure storage.

Types of Logbooks

Equipment Preparation, Usage, and Maintenance Logs

Equipment log books should be available for all major equipment and systems in a GMP/QSR production facility—including boilers, water for injection (WFI) stills, purified water systems, mixers, scales, pumps, tanks, packaging equipment, cell culturing machines, fill lines, lyophilizers, and sterilizers.

Production equipment logbooks for drug and biologic manufacturing must list the batch numbers of product processed by the equipment. This requirement is specifically outlined in 21 CFR 211.182:

> "A written record of major equipment cleaning, maintenance… and use shall
> be included in individual equipment logs that show the date, time, product, and
> lot number of each batch processed. If equipment is dedicated to manufacture
> of one product, then individual equipment logs are not required, provided that
> lots or batches of such product follow in numerical order and are manufactured
> in numerical sequence. In cases where dedicated equipment is employed, the
> records of cleaning, maintenance and use shall be part of the batch record. The
> persons performing and double-checking the cleaning and maintenance shall
> date and sign or initial the log indicating that the work was performed. Entries
> in the log shall be in chronological order."

Room Cleaning and Usage Logs

In some cases a room usage logbook replaces an equipment logbook. When a controlled area of the facility is dedicated to a certain activity and supported by specific and dedicated equipment, room use and equipment use become synonymous. For example, a vial labeling operation may always occur in a dedicated room where one logbook can be both a room and equipment log for recording room cleaning, equipment maintenance, and batch labeling activities.

Some areas with no major equipment installations require room cleaning and use logbooks. These rooms are usually controlled access areas—often clean rooms. A room used for

Master Cell Bank work, for example, requires strict control over cleaning and usage. No two cell lines may be open in the room at the same time, and cleaning and monitoring must occur between cell-handling events. All of these activities are recorded in the room cleaning and use logbook. Before technicians enter a room to work with a new cell line, they must record this event in the logbook.

Material/Information Movement Logs

The movement of materials, product, and information through processing areas occurs in all areas of the facility and the chronology of these events is recorded in logbooks. Logbooks are expected for:

Receiving in Material Handling

Shipping in Material Handling

Correspondence with FDA in Regulatory Affairs

Deviation Reporting in QA

Corrective Action Implementation/Follow-Up in QA

Document Change Orders in QA

Sample Receipt in QC

Notification of Results in QC

Adverse Event/Complaint Logs in Clinical Affairs, Marketing, or QA

Visitor Logs for the Facility at Reception

7

Quality Manuals

A QUALITY MANUAL IS A DOCUMENT or a compilation of documents that summarizes the corporate approach to quality commitments; it provides a menu to the document system that supports these commitments routinely. The quality manual concept was first introduced by the ISO 9000 standards and adopted for use by FDA in 21 CFR 820 in the 1990s. The quality system approach to compliance, summarized in a Quality Manual has since been adopted by many different types of industries and services, and throughout the world, providing a common language for quality expectations with vendors, contractors, and suppliers. In spite of the requirement for a Quality Manual in some industries, it is used in many more industries as a business control tool.

This Quality Manual provides a convenient way to define the organization, to designate business expectations and compliance requirements across functional areas, sites or contractors, and to act as a menu for all associated plans, protocols, procedures, and specifications. It is the "top document" in the document system.

Quality Manuals should be designed and written to provide a consistent approach to decision making at all levels of the organization. They should provide an overview of the operations to auditors and others without compromising proprietary information. They should establish specific organizational roles, relationships and responsibilities for those who work in or with the organization, and they should assign responsibility. They can also provide the basis for site or divisional harmonization by establishing the minimum standards of practice of the organization. ISO 10013, "Guidelines for Developing Quality Manuals" can be used as a reference.

In this chapter, example outlines of a Quality Manual are provided in Exhibits D, E. This example Quality Manual is a stand-alone document that should rarely change. It can be supported by a series of Standard Practice Documents. A Standard Practice Document (see Exhibit F) is a more function-specific document, which is expected to establish the minimum standards of the organization and assign responsibility for that function at a corporate level. Consult Chapter 9 in *Risk Management Basics*[2] for information about defining functional areas.

Tofte Medical, Inc.

Quality Manual Table of Contents

Exhibit D

Tofte Medical, Inc.   Page 2 of 2

Quality Manual Table of Contents

Tofte Medical, Inc.   Page 1 of 6

Quality System Manual; Version 1

1.0 Scope/Purpose of Quality Manual

This Quality Manual presents the Tofte Medical, Inc. (TMI) approach to organization, roles and responsibilities, and minimum standards of practice. The entirety of these commitments are called the TMI Quality System. As such, this Quality Manual presents or refers to:

- the basic components of the quality system;
- commitments associated with each component;
- the organization and interrelationship of these commitments;
- a menu of quality system documents that can be accessed to assure compliance; and
- an approach to management of these quality system components that assures customer and patient satisfaction, employee responsibility toward quality commitments, and shareholder support.

The Quality System is designed to ensure that

- products meet specifications/requirements for market;
- processes used to develop, produce, test, and distribute those products are consistent and reliable;
- the quality of the resources that support processing (materials, information, equipment, people, environments) is established and maintained.

In addition, this quality system is managed in a manner that assures ongoing notification of management about the analysis of quality system measures and any actions taken to correct real problems, prevent potential problems and facilitate product, process, and quality system improvements.

Each quality system component is defined in an TMI Standard Practice Document (SPD). These documents contain the commitments of the corporation to a given quality system component and any associated regulatory requirements (FDA 21 CFR 820, 21 CFR 600, 21 CFR 211; ISO 9001, appropriate ENs, and MDD 93/42/ECC Annex II), they assign responsibility to a functional area of the corporation, and they provide the road map to directive documents that ensure that a record of compliance is maintained continuously. Each SPD covers a logically separable part of the quality system.

This quality system applies to all TMI operations. TMI develops, manufactures, tests, distributes, and markets *in vitro* diagnostic products for the U.S., European, and international markets. Operations for TMI, Inc., are located at 94 SW Hwy 44 in Tofte, Minnesota.

Exhibit E

Tofte Medical, Inc.

Quality System Manual; Version 1

2.0 Quality System (QS) Management:
Commitment, Organization, and Responsibilities/Authorities

2.1 Corporate Mission and Letter from President (provide this here)

2.2 Organization and Interrelationship of Quality System Components

2.2.1 QS Components, Commitments and Responsibilities

TMI Quality System components have been identified from the FDA's 21 CFR 820, the ISO9001:1995 and ISO9004:1995, (as well as from ISO 9001:2000 and ISO17025) and the functional requirements of the TMI business operations.

Note: Each QS component is listed with its associated Standard Practice Document (SPD). Current quality system components include:

- Resource Management Practices
  Material Management Practices (SPD-01)
  Personnel Management Practices (SPD-02)
  Equipment Management Practices (SPD-03)
  Information Management Practices
      Documents (SPD-04), (SPD-05)
      Software (SPD-06)
      Contract Review (SPD-07)
      Regulatory Control (SPD-08)
  Environment Management (SPD-09)
  Facility Management (SPD-10)
  Finance (SPD-11)

- Product Development Management Practices
  Clinical Study/Design Validation (SPD-12)
  Process Development (SPD-13)
  Method Development (SPD-14)

- Validation and Transfer Management Practices
  Process Validation (SPD-15)
  Test Method Validation (SPD-16)
  Contractor Qualification (SPD-17)

- Product Manufacturing Management Practices
  Resource Processing (SPD-18)
  Manufacturing Operations (SPD-19)
  Testing and Inspection Operations (SPD-20)
  Product Dispositioning (SPD-21)

Exhibit E

Tofte Medical, Inc.

Quality System Manual; Version 1

- Product Management Practices
  Product ID, Traceability, and Inventory Control (SPD-22)
  Shipping and Distribution Operations (SPD-23)
  Product Stability Monitoring (SPD-24)
  Post-Market Surveillance (SPD-25)
  Record Management (SPD-26)

- Quality System Measurement, Analysis, and Improvement Practices
  Deviation, Unexpected, and OOS Result Management (SPD-27)
  Failure Investigation and Action Identification (SPD-28)
  Trending and Evaluation of Data for Potential Problems (SPD-29)
  Action Implementation and Effectiveness Follow-Up (SPD-30)

- Management Oversight Responsibilities for Quality System (SPD-31)

2.2.2 Interrelationship of QS Components

Resources ⅠⅠⅠⅠ➤ Processing ⅠⅠⅠⅠ➤ Product

TMI's quality system is designed to ensure the quality of its **products** by ensuring the quality of the **processes** that are used to develop, manufacture, test and distribute products, and to ensure the quality of the **resources** that support these processes. To ensure a consistent and committed approach, each type of resource and each type of processing have been assigned to a specific functional area of the corporation. That functional area defines its commitment to the quality system, from the perspective of a particular quality system component, in a Standard Practice Document.

In addition it is recognized as appropriate that for each resource, process or product, there is a lifecycle of development—validation—routine use—monitoring/ change that must be controlled. This lifecycle is also considered in the design of the quality system component identification and functional area assignments.

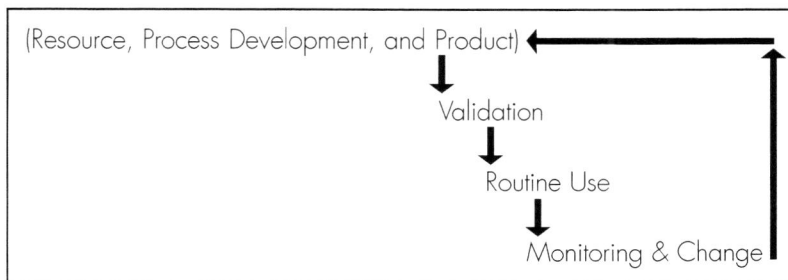

(Resource, Process Development, and Product)
→ Validation
→ Routine Use
→ Monitoring & Change

Quality System Manual; Version 1

2.3 Management Representative

The Management Representative is responsible for reporting the results of management review meetings to the President. The Management Representative monitors the work conducted by the quality system measurement, analysis and improvement functions, and reports to management periodically and on an as-needed basis.

3.0 Resource Management

The quality of the resources that support product processing at TMI are established in specifications, job descriptions, etc. and these resources are controlled (audited, tested, inspected) to ensure that the quality standards are met routinely. Resources include:

- Materials (purchased or prepared materials and processed components)
- Personnel
- Equipment (production, facility, testing, storage and distribution)
- Information (software, documents, contracts, regulatory commitments)
- Facilities and Environments
- Money

3.1 Material Management

Quality System Component Commitment Document = SPD-01

Responsible Functional Area = Material Handling

Commitments designed to comply with the following regulations

21 CFR 820.50, 820.60, 820.80, 820.140, 820.150, 820.160.
EN46001:1996 section 4.10.2, 4.15.2
MDD 93/42/EEC Annex II

General Topics of Commitment/Control

3.1.1 Purchasing Controls (SOP xx)
3.1.2 Receiving, Inspection, and Status Determination (SOP xx)
3.1.3 Inventory Control and Warehousing (SOP xx)

Tofte Medical, Inc.

Quality System Manual; Version 1

3.2 Personnel Management

Quality System Component Commitment Document = SPD-02

Responsible Functional Area = Human Resources

Commitments designed to comply with the following regulations
21 CFR 820.25
EN46001:1996 section 4.18
MDD 93/42/EEC Annex II

General Topics of Commitment/Control
3.2.1 Functional Area Roles and Responsibilities
3.2.2 Training, Qualification, and Assessment of Employees

3.3 Equipment Management

Quality System Component Commitment Document = SPD-03

Responsible Functional Area = Engineering

Commitments designed to comply with the following regulations
21 CFR 820.70, 820.75
EN46001:1996 section 4.11
MDD 93/42/EEC Annex II

General Topics of Commitment/Control
3.3.1 Procurement
3.3.2 Routine Operation and Use of Equipment
3.3.3 Preventive Maintenance, Calibration, and Repair
3.3.4 Qualification of Equipment

Exhibit E

Tofte Medical, Inc.   Page 6 of 6

Quality System Manual; Version 1

3.4 Information Management

The information required by TMI to reliably manufacture, test, and distribute products, as promised in regulatory submissions and contracts, is controlled. Information is managed and controlled as appropriate to the expertise in the associated functional areas.

Documents—Quality Assurance/Document Control
Software—Quality Assurance/Data Management
Contracts—Quality Assurance
Regulatory Submissions and Communications—Regulatory Affairs

Information and changes to this information over time are integrated into Operations through the Risk Management system and associated document change control procedures. Quality Assurance provides oversight to this information management function.

Note: The following sections would be formatted as 3.41—3.44 above

3.4.1 Documents
3.4.2 Software
3.4.3 Contract Review
3.4.4 Regulatory Control

3.5 Environment Management

3.6 Facility Management

3.7 Finance

Note: The remaining Quality Manual sections are completed as described in the Table of Contents.

Exhibit E

Tofte Medical, Inc.   Page 1 of 3

Standard Practice Document: Material/Product Management Practices

1.0 Purpose/Scope

This Standard Practice Document establishes the commitments for managing:

- purchasing, receipt, inspection/test, identification, storage, inventory control, movement, and use of materials;
- identification, storage, inventory control, shipment, and tracking of products;
- service contracts for testing, manufacturing, calibration, etc.

2.0 Commitments to Material/Product Management Practices

TMI's commitments to material, product and service management include:

- All items purchased for use in the production and testing of TMI products are purchased from vendors that have successfully completed a vendor qualification process.
- All services contracted to support the production, testing, and distribution of TMI products are contracted from organizations that have been evaluated and found acceptable to provide such services.
- All items used in the production and testing of TMI products are completely identified with part numbers, control numbers (lot numbers), and status indicators (released for use).
- All items received into the facility are recorded and evaluated (inspected/tested) to ensure that they meet pre-established specifications for quality, as directed.
- Movement of materials and product within the facility and to/from outside contractors is controlled (recorded/documented).
- All production events for component or final product production are scheduled to ensure that the proper resources are available to complete the work.
- All products in the facility, in all stages of production, are identified completely (part numbers, lot numbers, and status indicators) and products will be segregated in storage based on their status of quarantine, released or nonconforming/rejected.
- Warehousing practices ensure:
  - clean, orderly, and secure storage conditions;
  - controlled and appropriate environmental conditions of storage segregated storage of materials/products;
  - inventory control systems for materials and products.

Exhibit F

Tofte Medical, Inc.   Page 2 of 3

Standard Practice Document: Material/Product Management Practices

- All product distribution is conducted in a manner that ensures a record of accountability and traceability.
- Records that provide evidence for ongoing control/management of materials and product are retained, e.g.,
  - inventory control records,
  - product distribution records,
  - scheduling records,
  - specification testing records,
  - warehouse cleaning and environmental control records.

3.0 Responsibilities:

Materials Management is responsible for the design and implementation of the materials and product management system presented in this Standard Practice Document, and establishing associated procedures.

4.0 References

- 21 CFR 820.50 (a)   Purchasing Control: Evaluation of Suppliers
- 21 CFR 820.50 (b)   Purchasing Control: Purchasing Data
- 21 CFR 820.60   Identification of Product
- 21 CFR 820.65   Traceability of Product
- 21 CFR 820.70 (h)   Production & Process Control: Manufacturing Material
- 21 CFR 820.80 (b)   Receiving Acceptance Activities
- 21 CFR 820.90   Nonconforming Product
- 21 CFR 820.120 (b)   Labeling Inspection
- 21 CFR 820.120 (c)   Labeling Storage
- 21 CFR 820.140   Handling
- 21 CFR 820.150   Storage
- 21 CFR 820.160   Distribution
- 21 CFR 211.80   Control of Components: General Requirements
- 21 CFR 211.82   Receipt and Storage of Untested…
- 21 CFR 211.86   Use of Approved Components…
- 21 CFR 211.89   Rejected Components
- 21 CFR 211.142   Warehousing Practices

Exhibit F

Tofte Medical, Inc.   Page 3 of 3

Standard Practice Document: Material/Product Management Practices

- 21 CFR 211.150   Distribution Procedures
- 21 CFR 211.184   Component, Container,
                          Closure & Labeling Records
- 21 CFR 211.204   Returned Drug Product
- ISO9001/EN46001   4.3 Contract Review
- ISO9001/EN46001   4.6 Purchasing
- ISO9001/EN46001   4.8 Product ID and Traceability
- ISO9001/EN46001   4.10 Inspection and Test
- ISO9001/EN46001   4.12 Inspection and Test Status
- ISO9001/EN46001   4.13 Control of Nonconforming Product
- ISO9001/EN46001   4.15 Handling, Storage, and Delivery
- ISO13485; 4.6.3   Purchasing Controls
- ISO13485; 4.8 a,b   Product Identification and Traceability
- ISO13485; 4.15   Packing, Storage, Delivery

5.0 Definitions

Materials—Components, chemicals, subassemblies, and printed materials that are used to manufacture products.

Products—Products that are available for commercial distribution from TMI.

6.0 System Design Features and Processes

Note: Describe specific procedures that fulfill these commitments within TMI.

6.1 Purchasing Controls

6.2 Production Planning/Scheduling

6.3 Receipt of Incoming Materials

6.4 Identification and Inspection/Test of Incoming Materials

6.5 Movement of Materials

6.6 Material/Product Storage Practices

6.7 Control of Inventory

6.8 Distribution of Final Product

6.9 Product Identification and Traceability

6.10 Contract Review and Monitoring

Exhibit F

Section III

Functional Area Documentation Basics

8

Material Handling Documentation Basics

THE MATERIAL HANDLING FUNCTION ensures the identity of materials and products in all stages of processing; this is a fundamental requirement of Good Manufacturing Practices and Quality System Regulations. The Material Handling Department is responsible for establishing the standards for material and product control and for implementing these standards in routine practices. Procedures should be written to establish standards of practice in:

- purchasing (chemicals, components, labeling, equipment, etc.)
- vendor/supplier relationships
- production scheduling
- material identification (ID #, control #, and status indicators)
- flow of materials (documented movement) to/from Production
- flow of information about materials, product intermediates or product status to/from QC
- controlled, secure storage of materials, product intermediates, and products
- inventory control of materials, product intermediates, and products
- controlled movement of materials, product intermediates, and products
- product distribution

Material Handling Resources

Material Handling resources include:

- physical storage spaces that allow for controlled and secure storage of materials product intermediates, and product of varying status;

• computerized control systems for inventory control, distribution control and production scheduling or supply chain control; and

• trained material handling operators.

The most significant resource of Material Handling is warehouse space. Space issues are compounded by variable storage conditions for materials (2-8°C refrigerators, -20°C freezers, -80°C freezers, room temperature, flammables, dry and secure storage for labels) and associated status designations for all storage conditions to satisfy segregation of materials, product intermediates and products that are Quarantined, Released, On Hold, or Rejected. Office and warehouse space should facilitate the orderly flow of materials through the department from the receiving area through Production and finally back to the shipping area. Avoid any material flows where released materials move along a route that intersects with similar, untested, or quarantined materials.

Material Handling Practices and Procedures

Purchasing and Vendor Relationships

In a small company purchasing is often assigned to the same individual who is responsible for Material Handling. Purchasing of controlled items in a GMP/QSR manufacturer is expected to be a controlled process. There should be designated, approved vendors for items. Certification/qualification records for vendors must be on file before the items can be used in commercial manufacturing. There should be procedures on how to order items. The purpose of this controlled process is to increase the likelihood that materials brought into and used in the facility have consistent quality, since quality can be compromised by inconsistency in production sources, shipping, or handling.

Material Identification

Material Handling personnel are responsible for identifying incoming materials (incoming from vendors or incoming from Production). Material identification requires that items be tagged, labeled, or bar-coded. These identifiers are controlled documents. There are several types of identifiers or levels of identity that Material Handling must provide:

• Material Identifiers/Designations contain the following information:

– material/product identification #

– receiving code or in-house lot #

– quantity of material

– storage conditions

– expiration date (assigned by Material Handling from the specification document)

• Material Status Identifiers are available to provide the following designations:

– quarantined materials/product

– released materials/product

– rejected materials/product

– materials/product on hold, pending further processing or results of investigation

- Material Sampling Identifiers that are applied to drums or boxes that have been sampled.
- Material Partial Identifiers are applied to partially used inventory, either for sampling or for production use.

It is customary, when designing visible labeling, to have the following colors associated with the status designations: quarantine = orange, released = green, holding = yellow. These color designations are not mandatory, only customary. The importance of color, however, is to provide visual cues in the routine production environment that can help Material Handling and Production personnel maintain the segregation of similar-looking materials in different stages of processing. Since bar-coding practices can often eliminate these visual cues from the production environment, it is instructive to consider the human factors element in the design of material identity and movement. Support routine practices with useful labeling whenever possible.

Do not apply the identifiers in distant locations from one another on the drum or package. It is common practice to place identifiers together on a package, making the flow and change of status traceable. Quarantine identifiers are usually the first label applied to a material upon receipt. They provide both the identity and status information (see Exhibit G). When the material has been inspected or tested against its specification and released for use, the Release label is applied partially covering the quarantine label.

Sampling identifiers are associated with Partial identifiers, etc. When identifying materials in drums or covered containers, the identifiers must be placed on the outside of containers, not on the lids. Lids are easily interchangeable and could cause a mix-up of materials if interchanged. Labeling the outside of boxes or pallets, although common, can also lead to misidentified or unidentified materials when unpacking begins. Establish procedures for these circumstances.

Release label stock must be unavailable to Material Handling personnel; similarly the ability to release materials electronically must not be accessible to Material Handling personnel. These identifiers or this authority must be provided only by Quality Control after materials have been tested and found acceptable.

Bar-code systems for the identification of materials is now common. These systems provide for the electronic segregation of materials and product based on their identity and status. These systems certainly simplify some of the Material Handling requirements of warehousing because they can eliminate or minimize movement of materials, but they can also lead to a false sense of security. First, these systems *must* be validated. Secondly, perform a human-factors-based risk assessment for the flow of materials/product in the facility to ensure that "what can go wrong" is not likely to occur.

Material Receipt and Material Receiving Logbook

The movement of materials and product intermediates into the facility is recorded into a Receiving Logbook/Database. The logbook maintains information about the item and can be used to assign receiving codes or in-house lot numbers. Consider the following entries:

• date received

• receiving code or in-house lot number assigned

• item description

• part number

• manufacturer and/or distributor

• manufacturer's lot number

• # of items received and configuration

• date items labeled and moved to Quarantined storage

Inspection, Labeling, and Storage of Materials

When a shipment of materials is received into the facility, it should be checked to ensure that it is, in fact, what was ordered and checked visually for any obvious damage. There must be evidence for this inspection; this evidence can be collected on a form or in the receiving logbook. When materials do not pass this inspection, they should be moved into a Holding Area to await further investigation or processing. All materials/product must be stored and identified at all times. If it is not possible to complete the inspection and receipt of materials by the end of the day, do not leave materials in the receiving area. Move them to an On Hold status.

As materials are processed according to the instructions provided on material or product specification documents, their storage locations and status will change. Materials may move directly to the warehouse (if no QC testing is required), or to the Quarantine area. Certificates of Analysis and paperwork received with a shipment should be labeled with the item ID # and associated receiving codes. Original copies of these shipping documents should accompany the testing request forms to Quality Control.

Movement of Materials and Inventory Control

The flow of and movement of materials to and from the Material Handling Department must assure:

• accountability and traceability of materials;

• that items are clearly and completely identified;

• that materials are protected from damage, deterioration or cross contamination from other products, microbes, dust or toxins;

• segregation of acceptable materials from untested or rejected materials;

• segregation of materials of similar identity in different stages of processing.

The accountability of materials is ultimately documented in the inventory control system. Day to day, however, the documentation of this movement must be captured in a manner that does not interrupt the flow of work. A Move Ticket or some similar type of document or electronic entry can facilitate these events. A Move Ticket accompanies any item when it is moved from one department to another, i.e., when materials are moved from Material Handling to Production for an assembly event, when product samples are moved from Production to Quality Control for testing, or when finished product is moved from Production to Material Handling. The Move Ticket is a simple document, often designed as a prenumbered form with several replicate pages to facilitate reference, use, and filing requirements. It contains the raw data about the movement and use of materials, which is used to update the material control/inventory systems.

An inventory control system must be able to account for and track (a) the movement and use of materials from receipt to use, and (b) the movement of product intermediates and product from creation to use or shipment. Inventory control systems are usually software-driven electronic databases. Procedures must be written to support the use of these systems and validation must be performed before the systems are used routinely.

Warehousing Practices

Materials must be stored and transported in a manner that preserves their identity and protects them from damage or deterioration. Materials that are similar in appearance or color, should be segregated from one another to prevent mix-ups. Identification and status systems alone are not always sufficient to prevent a mix-up of materials. Product identifiers of similar size and color, for example, should be physically segregated from one another in the label storage areas and in the product shipping areas. One must ensure that hazardous materials are stored in a segregated manner that ultimately ensures the safety of the worker and the facility. Provide procedures to ensure these segregation practices.

All storage areas and chambers need to be completely and visibly identified. Items sitting on warehouse shelves must be visibly identified; status of materials/products must be evident either visually or by virtue of storage location (when bar codes are used). The area itself, whether a cold room or a warehouse area requires visible identity for status of materials and access security from every entry point. There should be no area of the Material Handling Department where the identity and status of materials is unknown. Identifiers that indicate the status of the items (i.e. Quarantine or Released), as well as safety identifiers for flammables, radiation, and other hazards must be clearly visible. Acids, bases, and organic solvents require special handling. Compressed gas cylinders must be identified and secured at all times. Provide procedures to assure these identification practices.

All pallets and all storage shelving should be off the floor and separated far enough from one another to permit thorough cleaning of the area. A policy should be established (in writing) about the rotation of stock in the warehouse. Typically a first-in-first-out (FIFO) system of control is sufficient. There should also be some provision for periodic review of materials for outdated stock. Provide procedures to establish these practices.

All quarantine, hold, and reject areas need to be secure. Controlled access should only be granted to QC and Material Handlers—not to Production personnel.

The controlled storage areas should be clean, free from rodents and insects, and environmentally conditioned to meet environmental specifications. These storage areas and chambers must be monitored routinely and records created that provide evidence of compliance to specifications. Logbooks for controlled environments are common, detailing routine cleaning, testing, and monitoring of these areas.

Production Scheduling

Production scheduling could also be managed by Material Handling personnel, since they maintain the information that supports supply-and-demand issues and decision making. Scheduling can be a simple task in a single product company, or a complex system requiring software controlled support in a multinational, multiproduct facility. No matter what the rigor of scheduling, however, the scheduling procedures should be established in writing, because errors in scheduling can expose materials and product intermediates to storage times and conditions that have not been validated. Unexpected delays in manufacturing can compromise product quality.

Distribution Control Practices and Shipping Logbooks/Databases

When final product is shipped to a customer or distributor, the movement of that item *from* the inventory control system is recorded. When that product enters the distribution system, it needs to be assigned a new set of identifiers to ensure accountability and traceability in the marketplace. Just as the receiving event initiates the inventory control of that item within the facility, the shipping event initiates distribution control of the product. Similarly, just as the inventory control system relies on the Move Ticket to gather the information it requires, the product distribution control system require a mechanism to collect the data on day-to-day movement of materials.

Product distribution must be recorded; in addition shipping of product intermediates to subcontractors, shipping of samples to laboratories, etc. should be recorded. A Shipping Logbook or Database is an appropriate and sufficient record. Similar to the Receiving Logbook, there can be a shipping code assigned to each individual shipment event. The Shipping Logbook records should contain:
- shipping date
- shipping identity designator
- the item part number or description
- item lot number
- number of units shipped
- special handling requirements (i.e., refrigerated shipment)
- destination
- shipping carrier and any carrier identification codes
- shipping confirmation designator

A record of the shipping event is a minimal requirement. For many medical products, however, accountability and traceability of the product to the patient is also required. In these cases, the product part number or catalog number, and often a serial number or a lot number is used to track usage.

Returned Goods

The accountability and traceability of returned goods must be carefully controlled and documented. Typically a Returned Goods Logbook is established and a returned goods numbering system instituted to track the materials and the investigation that supports their ultimate disposition. Returned goods must be stored in a secure area, separate from other product, until disposition is determined. They should be labeled with their returned goods identification number and a Reject, On Hold, or Quarantine identifier. The Returned Goods Logbook entries include:

- date of receipt
- a returned goods identifier
- product part number or catalog number
- product lot number or serial number
- customer who returned product
- original shipping code (if available)

It is good practice to complete a Returned Goods form that is used to notify Quality Assurance of the arrival of the items. If there has been a complaint in association with the return, QA might already know about the shipment. Often, however, returned product can arrive at the facility with no warning and the Material Handling Department is the "first to know."

Files in Material Handling

Records created to support the routine operations in Material Handling include vendor files, shipping carrier files, and distributor files. These files provide a history of the relationships, problems, complaints, etc. In addition, there is the need to create files for the records associated with the daily operations of:

- materials ordered but yet to be received;
- materials received and awaiting disposition;
- materials in hold areas, awaiting investigation;
- materials returned, awaiting disposition; and
- materials in area awaiting destruction, etc.

Procedures for Material Handling

Purchasing Standards and Procedures

Material Handling and Movement Practices

Warehousing Security and Practices

Cleaning of Warehouse Areas

Pest and Insect Control in Warehouse Areas

Control of Label Inventory

Standards for Packaging and Shipping

Material Handling Training

Inventory Control Practices

Distribution Control Practices

Receipt of Incoming Materials

Shipment of Product Intermediates and Final Product

How to Assign Receiving Code Numbers

Export Procedures

Handling of Hazardous Materials

Handling of Returned Goods

Handling of Rejected Materials

Operation of...equipment

Monitoring Controlled Storage Areas for Environmental Conditions

Label Production (if material identifiers are produced in-house)

How to Identify Distributed Products to Support Recall or Market Withdrawal Actions

Deviation and Unexpected Event Reporting in MH

Out-of-Specification Result Reporting in MH

Preliminary Investigations in MH

Records Created in Material Handling

Purchasing Records

Receipt of Incoming Materials Logbook Records

Returned Goods Records

Inventory Control Records

Shipping/Distribution Forms and Records

Material Destruction Logbook

Vendor/Supplier Audit and History Records

Distributor Audit and History Records

Tofte Medical, Inc.

QUARANTINED MATERIALS

Part # _____ RC # _____

Qty. _____ Configuration _____ of ____

Storage Conditions _____

Labeled by _____ Date _____

Orange

Tofte Medical, Inc.

RELEASED MATERIALS

Part # _____ RC # _____ Configuration ____ of ____

Expiration Date _____

Labeled by _____ Date _____

Green

Tofte Medical, Inc.

REJECTED MATERIALS

Part # _____ RC # _____ Configuration ____ of ____

Disposition Date _____

Labeled by _____ Date _____

Red

Exhibit G

9

Maintenance Documentation Basics

EQUIPMENT CONTROL IS A FUNDAMENTAL REQUIREMENT of GMP/QSRs. Consistent equipment performance supports consistent process performance, whether the process is a manufacturing process or a test method. Consistent equipment performance is achieved through appropriate and consistent equipment operation, maintenance, and control. The Maintenance Department should be the source of expertise for equipment, establishing the standards of equipment installation, operation, performance, maintenance, calibration, and repair. These standards of practice are established in Maintenance Department procedures.

Maintenance Department Resources

When establishing basic documentation for Maintenance:

- make an equipment list;
- identify the equipment-associated responsibilities of Maintenance (operation, cleaning, calibration, maintenance, etc.);
- ensure that SOPs are written and Maintenance technicians are trained in these procedures; and
- identify other Maintenance Department responsibilities and write supporting procedures for these duties, as well.

Consider that the following types of equipment require some level of identification and control to ensure their consistent performance:

- manufacturing equipment
- test equipment

- environmentally controlled chambers or areas (e.g., HEPA-filtered hoods, fume hoods, incubators, refrigerators, cold rooms, freezers, etc.)
- measuring equipment (e.g., calipers, gauges, pipettes, balances, stopwatches, timers, calculators)
- utility systems—boilers, water purification systems, water distribution systems, clean steam generators, compressed air, clean rooms, HVAC, waste treatment systems, and sprinkler systems
- calibration equipment and calibration standards
- forklifts and material handling equipment
- computerized data management systems
    - laboratory sample and data management systems
    - production scheduling computerized systems
    - inventory control computerized systems
    - distribution control computerized systems
    - computer hardware for database maintenance/entry
    - document control computerized systems
    - record archiving computerized systems

Additional responsibilities of the Maintenance Department might include:
- general facility and grounds upkeep
- building security
- building safety, (e.g., fire suppression equipment)
- operation and maintenance of utilities
- preventive maintenance programs for all equipment
- calibration programs for all equipment
- developing purchasing specifications for new equipment
- installation of new equipment and software
- installation qualification of equipment and software
- operational qualification of utilities and manufacturing equipment
- vendor and service contract relationships

Equipment maintenance and control activities can be handed down to other functional areas of the facility; Quality Control, for example, might want to perform routine maintenance and calibration of their equipment. When the authority to do the work is handed to another department, the responsibility for the work is retained in Maintenance. This means, in the example above, that Maintenance schedules the work, informs QC when to perform certain tasks and when the tasks have been performed, QC informs Maintenance to "close out" that task.

Facility Control Practices and Procedures

Maintenance is responsible for maintaining the design and control of the facility. Facility control practices include selection and maintenance of designated materials of construction, flow of people, flow of materials, flow of product, flow of utilities, general facility cleanliness, and utility system operation. Documentation basics associated with these responsibilities require that standards of practice be established in writing and that records be kept routinely to provide evidence that the standards are met.

Utilities

It is instructive to realize that the Maintenance Department is responsible for "manufacturing" many of the raw materials that the Production Department uses to make product. For example, Maintenance "produces" purified water, clean steam, clean room environments, and compressed air; these materials are used either directly in indirectly in the product. As a result, the quality of these utilities must be ensured just as they are ensured in Production.

The quality of the water, steam, air, etc., that is produced by a utility system is established in a specification document (see Chapter 4). Performance monitoring of these systems is required. Maintenance is responsible for monitoring equipment performance routinely. Monitoring expectations can be established in SOPs. Evidence of performance can be collected in logbooks or forms. For most utility systems, QC monitors and tests the "product" of the system routinely, and Maintenance monitors and tests the performance of the equipment.

Facility Upkeep and Security

Maintenance is usually responsible for the general upkeep of the facility and grounds. Even when these duties are contracted to outside services, they report to Maintenance. Duties include:

- cleaning and disinfection of general hallways, offices, bathrooms;
- insect and rodent control programs;
- waste disposal programs,
- grounds-keeping; and
- building and grounds security systems.

These programs must be implemented in the facility in a manner that does not compromise the quality or safety of the products. The types of cleaning agents, insecticides, etc., must be compatible with device or drug manufacturing environments and their usage must ensure that they do not come in contact with any product surfaces.

Security systems must be designed to protect not only the facility from outside intruders but also to protect "restricted access areas" of the facility from unauthorized entry by employees.

Equipment Control Practices and Procedures

Inconsistent equipment performance can lead to inconsistent product quality. Equipment performance can be controlled in many ways. The amount and the rigor of control is determined by the type of equipment, its complexity and its use in operations. Equipment performance commitments must be established in controlled documents and followed routinely. Equipment identification, installation, qualification, calibration, maintenance, and repair/ change offer opportunities for control that can affect consistent equipment performance.

Equipment Identification and Labeling

Equipment identification is required for *all* equipment used in the purchasing, development, manufacture, testing, and distribution of products. This ranges from major utility systems, such as boilers and water systems, to laboratory equipment, such as pipettes, calipers, and computer hardware. Equipment identification should be placed on the equipment in a location easily visible to users. This identity is used to link equipment performance in routine operations with equipment operation, maintenance and control records (see Chapter 1).

Equipment, however, it not always a free-standing unit of operation. Equipment systems are common in the industry, e.g., water purification and distribution systems, product manufacturing systems, and HPLC testing systems, which consist of numerous pieces of equipment that are linked together to perform a task. The need to identify and establish equipment configurations for equipment systems is fundamental to equipment control practices. Establish within the equipment ID # system a mechanism for identifying equipment units that are connected to one another. In the following example, the equipment identification number format would facilitate the identification of the associated equipment system:

> EM-PW-PP01 is the identifier for a pump in the Purified Water system,
> and
> EM-PW-FT14 is the identifier for a filter in the Purified Water system.
> > where E indicates that this is an equipment identifier
> > where M is a location identifier (MT = Maintenance)
> > where PW is an equipment or equipment system identifier
> > > (PW = purified water system; FT = filter)
> > where PP is the type of equipment (PP = pump)
> > where the digits are unique equipment identifiers.

Equipment Specifications/Equipment Master Cards

Equipment System Master Cards are documents that can be used to describe the equipment, its features, and characteristics. Master Cards are equivalent to a material specification document, in that they should provide all of the essential information about the equipment in a convenient location. The mechanics and technicians in the Maintenance Department should consider Master Equipment Cards as a primary resource of information about the equipment.

The Equipment Master Card can include:
- the equipment identification number
- date of installation, P.O. #

- any in-house asset numbers
- equipment manufacturer and/or supplier, model #, serial #
- specific location of the equipment
- equipment configuration
- spare parts, vendors and catalog numbers
- storage reference numbers for the in-house spare parts warehouse
- outside calibration services or outside repair services
- reference to SOPs and programs for calibration and maintenance
- reference to SOPs on operation or, cleaning of, etc.
- warning statements about working with the equipment
- a listing of operating and repair manuals for the equipment

Equipment Installation

The Maintenance Department is responsible for the installation of all new equipment; including installations performed by a contractors or vendors. Installation of major utility systems, requiring validation, must be documented in Installation Qualification documents. (See Chapter 13 and Exhibit J)

When Installation Qualification is not required, the equipment should be identified with an equipment identification number and a Master Equipment Card should be completed. When equipment has software-driven controlling or monitoring features, the software must also be identified and controlled, by assigning a material identifier to the software and controlling its revision through the document change control system.

Equipment Operation Procedures

Write standard operating procedures (SOPs) for equipment operation when equipment is new, preferably before start-up. Procedures that describe start-up, alarm reset, and shut-down of all major equipment systems, such as water systems, clean steam generators, and compressors are expected. In addition, write SOPs for routine maintenance operations, such as blow-down of boilers, filter changes, and regeneration of purified water system resin beds. A procedure is required when the tasks must be performed in a certain order to ensure the safety of the operator or the consistent performance of the equipment. Procedures do not need to be written for tasks that any Maintenance technician has been trained to perform.

Procedures that describe how to start up equipment, etc., do not simply restate the instructions in the equipment operation manual. Procedures in a GMP/QSR facility must also instruct the operator to record information and data observed during these tasks onto pre-established data collection forms or logbooks.

Equipment Usage and Area Usage Logbooks

A logbook is a chronological record of all equipment-related activities; it provides easy access to information about equipment status at any time. Equipment logbooks should be available for all major equipment and systems in a GMP production facility—including boilers, water production systems, water storage and distribution systems, packaging equipment, and sterilizers. When in use, it must be stored on or near the equipment or area where it is used. (Note: Plastic file holders are convenient for attaching the books to the machinery or a nearby wall.) When it is filled, it should be returned to the Document Department (see Chapter 6).

Equipment Preventive Maintenance and Calibration Programs

Once installed, equipment must be maintained to ensure reliable and consistent performance and to minimize malfunction and downtime. Maintenance must establish routine maintenance tasks, a schedule for them to be performed, and procedures to use when performing the tasks. This collection of tasks and the associated schedule is called the Preventive Maintenance Program. Consult *Documentation Practices*[1] for a description of program documents used to support monitoring and control programs such as preventive maintenance and calibration. The PM program can be organized manually or on a database. Whatever the method, there should be:

- a list of tasks and a commitment to their frequency of performance for each equipment unit;
- methods used to do the work, i.e., SOPs; and
- data to ensure that the tasks are performed in a proper and timely manner.

Equipment calibration commitments are managed like preventive maintenance commitments, requiring a list of tasks and the frequency of their performance for each equipment unit. Calibration events must be recorded on a form or in a logbook. In addition, a calibration identifier should be applied to the equipment which indicates the equipment ID #, date of calibration and calibration due date. Equipment that does not require calibration can be labeled "no calibration required."

Equipment Repair/Change

All unscheduled repair events must be documented. This document, often called a Maintenance Work Order, can serve as both the repair requisition form and the repair history record. When the repair is completed a copy of the form is filed in the Equipment History File, discussed below.

When critical utilities or equipment are shut down for repair, system qualification or validation may be required before the system can be used by or for production. In these cases, the Equipment Repair Work Order must be signed by QA or Validation before the equipment is recommissioned. Equipment that is not available for use because of repair or maintenance procedures should be identified as "locked out." Create a label or tag to identify this condition.

Equipment History Files

When equipment has been installed and assigned an identifier, an Equipment History File can be created. Equipment history files should be kept in a secure but accessible location within the Maintenance Department and should be considered a key Maintenance resource. These files contain reference information for the equipment as well as a history of installation, calibration, maintenance, change, and repair. These files could contain:

- The Equipment Master Card
- In-house procedures for the operation of the equipment, calibration, maintenance, etc.
- Records—filed chronologically—of all repair events
- Records—filed chronologically—of all calibration events
- Records—filed chronologically—of all preventive maintenance activities
- Installation Qualification Records
- Vendor service records
- Inspection by state, local, or certification authorities
- Equipment manuals and associated vendor information

Maintenance Procedures

Equipment Identification Numbering

Building Security and Response to Alarms

Calibration of…various equipment

Operation of…various equipment

Routine Maintenance of…various equipment

GMP Training of Maintenance Technicians

New Equipment Installation Practices

Power Failures: How to Restore Facility to Full Operation

Preventive Maintenance Program

Calibration Program

Equipment Repair Practices and Records

Deviation and Unexpected Event Reporting

Out-of-Specification Result Reporting

Preliminary Investigations in Maintenance

Safety Practices

Regeneration of Resin Beds

Rodent and Insect Control Programs

General Facility Cleaning and Sanitation Programs

Use of Contractors

Waste Disposal Procedures

Response to Equipment Alarms

Records Created by Maintenance

Preventive Maintenance Records

Calibration Records

Repair Records

Contract Servicing Records

Equipment Vendor Audit and History Records

Equipment Installation and Change Records

Equipment History Files

10

Quality Control Documentation Basics

THE QUALITY CONTROL FUNCTION is primarily responsible for the testing and inspection of:

- incoming materials, components, labeling, etc.
- processing intermediates/subassemblies
- final product
- environmental conditions in production, material handling, etc.
- utility system performance
- validation study samples
- stability study samples
- product and process development samples.

The size and complexity of the Quality Control Department will depend directly on the type of product, the complexity of the production process and the number of units required for market. A device manufacturer, for example, may or may not require major utility systems such as Water for Injection, clean room environments or steam-in-place systems. The product may or may not be easily assembled and packaged on site, may or may not require sterilization by a contract sterilization service, may or may not have a drug or biological component, and/or may or may not require unique or sophisticated testing methods. It is this mix of possibilities and concerns that will most directly affect the size and complexity of the Quality Control Department.

The purpose of the QC laboratory is to produce data; the "production lines" that support these operations are the test methods. The *resources* required to perform data production must be established and controlled in documents; the data production events (*test methods*) must be established and controlled in documents; and the controlled storage, handling, and transfer of the data product must be established in documents.

Resources for Quality Control

The quality of the resources required to perform QC analysis should be controlled and should be established in documents. Specification documents and procedures generally are sufficient for establishing these controls.

Laboratory Space

Laboratory space will vary according to the responsibilities of the QC department. Manufacturers of simple device products will require only tabletop stations for inspection and testing of medical device components/products. Manufacturers of biologic or drug products will likely subdivide the areas into Microbiology, Chemistry, Bio-assay, Inspection/Process Monitoring, etc.

The flow of samples into the laboratory and the flow of data out of the laboratory must be controlled. Samples must be completely identified and some mechanism (logbooks, databases or LIMS) of accountability and traceability instituted to ensure timely analysis. Similarly data must be reported in an efficient manner and raw data filed in a way that assures access and retrieval. Procedures should be written to support the operation of these systems.

Plan the laboratory to facilitate the flow of samples and product, the flow of clean and dirty glassware, the flow of waste, the flow of people, and the flow of data. Establish this flow of materials in written procedures that support these activities.

Laboratory Test Equipment

All test equipment must be properly installed, used, maintained and routinely calibrated. Equipment files that provide a history of installation, repair and calibration should be maintained either in the QC laboratory or in the Maintenance area. There must be evidence that calibration has occurred and is acceptable; calibration information attached to the equipment should indicate the equipment identification number, the date of calibration, and the date of recalibration. Calibration programs should include thermometers, balances, pipettes and calipers as well as the calibration of more sophisticated instrumentation (see Chapter 9).

Procedures should be written for the operation, cleaning, and calibration of equipment. In addition, equipment such as HPLC units, which might require cleaning and suitability testing between uses, should have an equipment logbook that records these events chronologically (see Chapter 6).

Laboratory Chemicals, Solutions, and Standards

The preparation of reagents and solutions in a GMP/QSR laboratory should be traceable to their original chemical components. This accountability and traceability can be maintained in a Solution Preparation Logbook, designating the solutions prepared, volume, concentration, chemical component identification, and date. Logbook entries can be supported with solution preparation forms for solutions requiring extensive formulation and testing.

All reagents and solutions should be completely identified with a part number, lot number, storage conditions, date of preparation, and expiration date. There should be procedures for periodic inspection of chemicals, standards and solutions to ensure that they have not expired or degraded during storage in the QC laboratory.

Trained Technicians

QC technicians must be trained to perform test methods. The content of training programs or expectations must be documented in SOPs and evidence that training has occurred recorded in forms or training logbooks.

Quality Control Practices and Procedures

Samples and Controlled Sampling Procedures

There should be procedures that describe sampling plans for components and product. These sampling plans must be based on sound rationale or some accepted sampling plan such as U.S. Military Standard 105E. Once samples are delivered to the laboratory, they should be completely identified, stored and used as directed, and destroyed when appropriate. Sample handling control practices should be established in written procedures.

Testing

Test methods are established in SOPs; in some companies there is a subset of procedures created specifically for test methods. Test method procedures are usually accompanied by forms that facilitate the collection of routine data from these testing events. This continues to be common practice, except for testing that accompanies final product release testing and stability study testing.

Increasingly, companies are creating what is called a Quality Test Record. These records are usually created to support the testing of finished products at release or during stability study. They are formatted like a Product Manufacturing Record in that they provide both the directives to perform the work and a fill-in-the-blank format for data collection on an event-by-event basis. The Quality Test Record is issued for the testing of a specific lot or batch of product, just like the Manufacturing Product Record. This helps to ensure the integrity of the record, minimizing the opportunity for a technician to destroy a form containing unacceptable data.

Data Management and Notification

Just as the flow of samples should be designated to assure that samples are analyzed in a timely manner, the creation and movement of information and data from the QC laboratory needs to be specified in procedures. Each test method should indicate how data is calculated and managed on an event-by-event basis. In addition, the laboratory should establish data management practices and data archives.

When testing is complete and the data is available, notification of results is an important aspect of QC responsibilities. Establish routine procedures for notification.

QC Procedures

Calibration and Maintenance of Laboratory Equipment

Cleaning and Disinfection of Laboratory Areas

Environmental Monitoring Procedures for Clean Room Areas

General Solution, Reagent, or Standard Preparation

GMP Training of Technicians

Handling Out-of-Specification Results

Deviation and Unexpected Event Reporting

Preliminary Investigations in the Laboratory

Handling of Hazardous Materials

Retesting Practices

Data Integrity Practices

Sampling Techniques and Documentation

Standards: Use, Control, and Storage

QC Testing of Materials/Product: Flow of Work

Operation of…equipment

Test Method SOPs

Utility System Monitoring Procedures

Stability Testing Monitoring Procedures

Records Generated by QC

Product Testing Records

Material/Component Testing Records

Stability Study Testing Records

Utility System and Environmental Monitoring Testing Records

Contract Laboratory Audit and History Records

Equipment Performance Monitoring Records

Technician Training Records

11

Quality Assurance Documentation Basics

THE QUALITY ASSURANCE FUNCTION assures that the quality commitments, established by the corporation and fulfilled by the various functional areas, are met. These commitments include:

- regulatory commitments as declared in product and facility submission documents;
- regulatory requirements such as Good Manufacturing Practices, Quality System Regulations, Good Clinical Practices, Hazardous Material Handling Practices, safety practices, etc.;
- commitments to clients in contracts and purchase orders; and
- good business practice commitments of the corporation.

The Regulatory Department, along with Product Development engineers/scientists, should provide Quality Assurance with the commitments and the standards against which the manufacturing process and the manufacturing facility will be judged. Quality Assurance, in turn, helps to establish the specifications, procedures, systems, and programs that will meet these requirements during routine operation.

Quality Assurance responsibilities include:

- release of product intermediates and final product
- design, implementation and maintenance of a document system
- design, implementation and maintenance of a record archiving system
- coordination and/or auditing of vendors, suppliers, and contractors
- coordination and/or auditing of internal operations for compliance with regulations and corporate documents

- training personnel in general principles of GMP/QSR operations and manage/coordinate the functional area-specific training
- review and approval of all documents
- helping to establish specifications for materials, products, vendors, suppliers, equipment systems, automated data management systems, etc.
- auditing of data monitoring programs for adverse trends in data of compliance[2]
- design and management of deviation reporting systems
- design and management of failure investigation meetings
- design and management of corrective/preventive action implementation/effectiveness programs
- interfacing with Regulatory Affairs to stay current on regulatory requirements
- interfacing with Validation to stay current on validation limits and controls
- design and management of the introduction of new/changed products and processes into existing operations

Quality Assurance Resources

The QA department requires office areas with ample counter space, filing cabinets and/or space for computer monitors and associated hardware to support document and record control practices. QA should be organized to facilitate a logical flow of work as documents and records are created, reviewed, approved, distributed, archived, and changed. Document and record archive areas should be restricted access areas.

Equipment requirements include numerous computerized data management systems; all of these systems will require compliance with 21 CFR 11. No matter what the status of the paperless factory, there continues to be a need for good copy machines that can collate and staple documents, a paper shredder, and filing cabinets. When training is a QA responsibility, training equipment may be required, such as video cameras and projectors for slides, computers, videos, and transparencies.

QA should be organized to support a logical flow of records during their review. Record archiving and storage areas should support the security and access needs of the facility. The confidentiality of records must be ensured through storage and filing practices. Without these records supporting the product in the marketplace, the corporation is at risk. There should be programs to store records in fireproof, off-site storage facilities, as well as programs for electronic storage of documents.

QA Practices and Procedures

Vendor Certification Audits

A vendor's Certificate of Analysis cannot be accepted until the reliability of the supplier's analysis has been established. This reliability can be established by testing vendor-supplied product to ensure that it meets vendor specifications. For materials and components whose quality could adversely affect the safety or quality of the final product, performing a vendor audit is also fundamental to Good Manufacturing Practices.

Ensuring the quality of products and services through auditing is often a responsibility of the Quality Assurance department. The purpose of auditing vendors, suppliers, and contract service organizations is to (a) assure, by inspection, the quality of their manufacturing or service organization, (b) assure, by inspection, that the characteristics of their products/services that most affect the safety and quality of the product are properly controlled and maintained, (c) minimize the burden of comprehensive QC testing of vendor materials upon receipt at the manufacturing site, and (d) establish a relationship that facilitates communication about problems and changes.

When establishing a vendor certification program, start by making a list of vendors and outside testing services. Determine which vendors supply materials or information considered critical to your product. Determine which vendors have an acceptable or unacceptable performance history. Determine which vendors might be making a unique item or a common item according to a unique or special process. With this information, assign auditing priority to the list of vendors.

Audit vendors for their ability to produce product in a manner that assures the consistency of product characteristics from lot to lot or unit to unit. When appropriate, audit vendors for compliance with Good Manufacturing Practices, ISO 9001 or basic quality assurance principles. Establish the criteria for vendor certification before performing the audits. Performing an audit is not, in itself, sufficient for certification; the vendor must pass the audit. Audit acceptance criteria is a fundamental requirement of certification that must be documented.

Maintain a vendor file system in either QA or Materials Handling that contains information about the company and its services, a listing of what items are currently purchased from this company, and the history of purchase, deviations or investigations of product and audit reports.

Internal Compliance Audits

Quality Assurance must also routinely perform audits of internal operations. The primary purpose of this audit activity is to ensure that work is being performed according to written commitments and procedures and that those commitments and procedures comply with external regulatory requirements. Consult *Risk Management Basics*[2] for detailed expectations for the management of audit findings.

Deviation Management and Failure Investigation

There should be programs and procedures in place to facilitate the documentation of deviations, unexpected results, and out-of-specification results in all departments. Investigations of these events and the rationale for actions resulting from the investigations must be documented. All investigation summaries should include a Quality Assurance review and signature. Given that QA has approval responsibilities for specification documents, they should be informed of all out-of-specification results.

There should be an failure investigation logbook to record these events, chronologically. The log should indicate the date of the event, the items affected, reference to where the original documents supporting the investigation can be found, and the date of resolution. The documentation supporting the investigation and resolution of the event should be kept with the documents affected by the investigation, such as the Device History Record/Production Batch Records or the specification testing forms.

Corrective/Preventive Action Implementation and Effectiveness Assessments

Actions authorized for implementation, as a result of an investigation, need to be managed consistently and tracked to ensure that they are effective. Procedures describing the action authorization and implementation program must be written. Forms, logbooks, etc. are used routinely to manage the information generated from this QA program. Consult *Risk Management Basics*[2] for more information on the design and content of these programs.

Dispositioning of Production Intermediates/Assemblies and Final Product

The release or rejection of final product is a primary responsibility of Quality Assurance. In a regulated industry, this release event must ensure that (a) the records that support product manufacturing and testing are complete and acceptable, and (b) final product meets the requirements of the associated specification document.

Note: It is instructive to realize that final product can meet final product specifications and still be unacceptable for the market. It is the responsibility of QA to review all records associated with the product and determine if there are any factors that could adversely affect the safety or efficacy of the product. Procedures for product dispositioning should be written to ensure consistent and complete record review.

Employee Training

All employees in the manufacturing facility should be trained in Good Manufacturing Practices and/or Quality System Regulations. This training should include job-specific and department-specific training, as well as facility-wide training in Good Manufacturing Practices, corporate policy and procedures.

In a small company QA usually coordinates the facility-wide training programs and ensures, through audit, that the departmental commitments to training are fulfilled. All training program content should be documented. In addition to training, there must be some evidence that the training was effective, which can be provided through classroom testing of employees and/or through documented observations of their work habits. Whatever the evaluation process, make sure that the evidence of acceptable employee performance is on file.

Document Systems

Documents are the tools of the Quality Assurance department. There are documents which describe the commitments of the corporation, documents which describe how those commitments will be fulfilled during routine operations, and documents which facilitate the collection of data. These documents must be properly reviewed, approved, and controlled.

The documents are not created by QA, they are, however, often managed by QA.

Document system procedures include:

- document creation, change, review, and approval
- routing of draft documents
- routing of approvable documents
- mastering of a document and distribution for use
- document files
- records and document shipment practices
- operation of the electronic document management system.

Access vs. Security of Records

A difficult issue to resolve is the need for full and convenient access to archived records *and* their security. Optical imaging of records may accommodate the need for both access and security, but until this equipment is in place, a makeshift approach is likely to evolve in response to this inherent conflict. It is common, for example, for supervisors to make unofficial copies of Product Manufacturing Records for their own reference. Although this facilitates their personal access to Manufacturing Records, the records that they copy are usually reproduced *before* final review and approval (i.e., before documentation process is complete). Incomplete records that are used as references are a liability both for the user and the corporation. If copies are useful, ensure that they are copies of fully reviewed and approved records and clearly marked as "copies"; also ensure that copies are destroyed in accordance with corporate document destruction policies.

QA Procedures

Contract Review Practices

Responsibilities of Quality Assurance

Expiration Dating Guidelines for Components and Materials

Employee Training Programs

Stability/Reliability Testing Guidelines

Auditing Practices

Reporting Deviations, OOSR, and Unexpected Events

How to Assign Document Identification Numbers and Versions

How to Assign Material Identification Numbers

Failure Investigations

Corrective and Preventive Action Identification and Authorization

Action Implementation and Effectiveness Assessments

Trending of Monitoring Data

Complaint Handling

Data Collection Conventions

Release of Final Product

Documentation Basics: Specifications

Documentation Basics: Standard Operating Procedures

Documentation Basics: Protocols

Documentation Basics: Data Collection Forms, Lists, Logbooks

Documentation Basics: Batch Records or Device History Files

Record Created by QA

Complaint records

Stability testing records

Product testing records

Material testing records

Failure investigation records

Deviation records

Corrective/Preventive action records

Contract monitoring records

Audit records

12

Production Documentation Basics

PRODUCTION IS THE FUNCTIONAL AREA that is responsible for all product-specific manufacturing events. "Manufacturing" is defined by the FDA to include formulation or assembly, labeling, packaging, inspection, and distribution. In biologic production and more recently in drug production, "manufacturing" also includes the production of source materials. Source materials for drug/biologic manufacturing are called "drug substance" or "active pharmaceutical/biologic ingredients" (APIs). Excipients are other ingredients in the formulation, not considered pharmacologically active. The Production department can also be responsible for the final preparation of materials used in processing (e.g., cleaning and sterilization of components, compounding of formulation reagents, dispensing, etc.).

Production Resources

As indicated throughout this text, all other functional areas within a corporation work to create the materials and information required for Production to be productive. Material Handling and QC produce released materials; Maintenance and QC produce acceptable utility resources (water, air, steam, environmentally controlled space); QA reproduces and issues the instructions for manufacturing; Maintenance and Validation provide the confidence in equipment and process performance. With this support system in place to provide resources, the Production Department makes products.

Production Practices and Procedures

Production activities can be separated into two distinct levels. There are the procedures associated with actual product manufacturing events (see Primary Manufacturing Processes), and there is a set of activities associated with the final processing of resources used in manufacturing or the activities associated with the close-out and clean-up of equipment and areas after product-specific manufacturing events (see Secondary Manufacturing Processes, page 89).

Although the Production Department has procedures, logbooks, forms, etc., just like any other department, this chapter will focus on the Product Manufacturing Record. The Product Manufacturing Record provides the detailed instructions required to make product.

Primary Manufacturing Processes

Bulk API Production or Reagent Manufacturing

Fermentation and Harvest

Formulation

Synthesis

Clarification and Concentration

Purification

Viral Inactivation

Filling

Lyophilization

Freezing

Bulk API Packaging

Holding and Storage

Labeling

Packaging

Testing

Device or Diagnostic Kit Assembly

Assembly of Functional Units

Functional Unit Cleaning/Sterilization

Component Labeling

Kit Assembly

Sealing

Sterilization

Labeling

Packaging

Testing

Drug/Biologic Final Product Production

Compounding/Formulation

Sterile Filtration

Filling

Stoppering

Vial Imprint Labeling

Sealing

Lyophilization

Terminal Sterilization

Inspection Labeling

Packaging

Testing

Holding and Storage

Shipping/Distribution

Secondary Manufacturing Processes

Area Preparation

    Cleaning and Disinfection

    Environmental Monitoring

    Inspection/Preclearance

Component Preparation

    Assembly of Nonfunctional Components

    Vial Washing and Sterilization

    Stopper Wash and Sterilization

Equipment Preparation

    Equipment Setup, Preparation, and Suitability Testing

    Equipment Cleaning/Sterilization

Solution Preparation

    Media Prep

    Buffer Prep

Cell Preparation and Scale-Up

Distinguishing primary from secondary manufacturing processes requires that the processing records designed to support primary manufacturing events are distinct from the processing records designed to support secondary manufacturing. Primary manufacturing records, traditionally known as Device History Records or Batch Production Records, provide a step-by-step account of product-specific processing. They are generated to support a lot/batch of product manufacturing in the same cycle of production.

The processing records designed to support Secondary Manufacturing Processes are not linked—on an event-by-event basis—to product lot/batch numbers because they are often used to support many different types of products. These processing events/records may be assigned lot numbers, or "date of processing" as control indicators, but these are *process*-specific lot numbers, not *product*-associated lot numbers.

The segregation of primary from secondary records facilitates the movement of materials and documents through the Production area. It provides for a clear separation of time- and order-sensitive production processes (primary manufacturing) from those production process-es that are not (secondary manufacturing). This can keep Production from committing resources to a batch or lot before the materials, equipment, and environmental conditions are acceptable. This segregation can also facilitate the triage of deviations and failures in pro-cessing. When failures in secondary manufacturing processes occur, and they are known before these materials are used in primary manufacturing events, the quality of the product is not compromised and Production can recover from the failure without compromising the records for other preparation events or the final product manufacturing record (because the product lot number is not associated with the secondary manufacturing events).

Segregating secondary from primary manufacturing events can also facilitate the triage required for corporate deviation monitoring and investigation systems, as failures in second-ary manufacturing can implicate several product lines and failures in primary manufacturing usually implicates one product line.[2]

Product-Specific Manufacturing Records

The product manufacturing record must:
  • be compliant with regulatory submission commitments;
  • be compliant with industry standards for GMPs/QSRs;
  • facilitate complete and accurate data entry practices;
  • complement a workflow on the floor that is achievable and will facilitate increased
    production capacity; and
  • incorporate validated limits/controls.

In addition, all product manufacturing records should:
  • reduce/minimize redundant or unnecessary data entry;
  • reduce/minimize signatures;
  • allow the record to be completed, logically and chronologically, as the work
    proceeds; and
  • facilitate a complete and timely review/approval of associated records.

All of the product-specific manufacturing events required to produce one lot or batch of product intermediate or product should be linked into one manufacturing record. It is rec-ommended, therefore, that paper-based batch records be segmented into major processing sections, with each section representing a distinct process that can be prepared, independent of other processes. For example, although aseptic filling and lyophilization might be process-es of the same manufacturing event, the preparation of the filling area/equipment and the preparation of the lyophilizer may need to occur simultaneously and in different areas of the facility. Segmenting the record to provide the lyophilization area with the directives it will need to prepare for the product as a separate, but linked set of instructions *and* providing the aseptic filling operators with a set of directives that they will need to prepare for the product serves the needs of the document user and facilitates a timely flow of product through the production area.

For example, an Enzyme Product Batch Record is assigned a distinct lot number for every batch produced. The actual Product Manufacturing Record, however, is segmented into 9 distinct sections that are stapled together individually. Each section is formatted the same way, as will be discussed below with preparation, processing and close-out events. The sections of this example record are:

```
A  =  Bulk Ingredient Thawing
B  =  Formulation
C  =  Sterile Filtration
D  =  Aseptic Filling and Stoppering
E  =  Lyophilization
F  =  Capping
G  =  Inspection
H  =  Labeling
I  =  Packaging
```

Each section of the record is formatted to include these fundamental elements of controlled processing:

Preparation and Process Clearance

This subsection provides for the preparation of all reagents, buffers, tanks, columns, filters, etc. required to do the work for *this* processing event only. Before moving from "preparation" to "product processing events," a Supervisor "clears" the event, meaning that they assure (with their signature) that the chemicals, equipment, processing area, and people are prepared and acceptable for use. The order of work in the "preparation" subsection is not critical.

Bill of Materials—The Bill of Materials (BOM) designates all required reagents, chemicals, components required to perform the processing in this section of the record only. It may be appropriate to create a BOM for materials and a BOM for equipment and component preparation events. From a GMP perspective, it is expected that a manufacturer will be able to identify all components used in the manufacture of a given lot of product. The Bill of Materials, in or attached to the product manufacturing record, is commonly used to establish this accountability/traceability. The Bill of Materials identifies product components, i.e., "components necessary for the preparation of the dosage form" (21CFR211.186(b)(4)). This is generally interpreted to mean

- product constituents (active ingredients, WFI and excipients),
- materials that are in contact with the product throughout its shelf-life (vials, stoppers, etc.), and
- materials that come in contact with the product during its processing (filters, column resins, buffers, reagents, etc.).

The Bill of Materials must also indicate the quantity of these components used in the production of a batch. The BOM is used as a checklist to ensure that all materials, equipment, etc. are acceptable for use before the work begins.

Preparation Directives—These are simple and brief entries that provide an opportunity to describe simple preparation events, to confirm that complex preparation events have been performed, and to link the product manufacturing record to the associated preparation records.

Processing Clearance

This activity serves as a check on the processing area before product is committed. This clearance provides an opportunity to assure that all GMP compliance requirements are met.

Product Processing Events

This subsection provides a detailed, step-by-step, chronological list of the tasks required to process product through the designated phase of product processing.

Section Close-Out

This subsection provides for an accountability and timely review of the work as product is moved to the next phase.

Reconciliation—A product batch record must also provide for the reconciliation of the components used for each production event. If 32,000 vials are moved into a filling area for use in a batch, one must account for (reconcile) the use of those 32,000 vials (vials for product use, vials returned to inventory, vials destroyed, etc.). Note that this is **not** a reconciliation of vials, it is a reconciliation of the vials *issued for use in this lot*. The reconciliation of each lot of vials is done through inventory control procedures in the warehouse. The GMP requirements for control of components and the records required are established in 21 CFR 211.184. In the end, however, from the perspective of the "batch" and its associated batch records, one must only identify what was used and how much. In identifying "what was used" (with a part number and a lot number), one should be able to "trace" from the batch record to associated component testing and preparation records.

There are limits associated with reconciliation, beyond which an investigation is initiated. If the reconciliation limit for vials in a 32,000 batch run is +/- 0.5% and one can only account for 31,000 vials, then an investigation is initiated.

Expected Yield—Expected yield is a requirement for major processing events in drug and biologic manufacturing (21 CFR 211.103). There should be an expected yield designated on the product manufacturing record, and the yield should be calculated for each batch produced. Expected yield is a process control, meaning that if the actual yield is not within the limits of the expected yield, then an investigation should be initiated. Do not confuse yield with reconciliation.

Acceptance Criteria—This is a list of information that must be fulfilled to consider this section of the record acceptable.

Attachments—List any attachments that *must* be a part of this section of the product manufacturing record when it is complete.

Consult Exhibit H for an example of a Labeling Record.

Secondary Manufacturing Preparation Records vs. Procedures and Forms

Secondary Manufacturing Preparation Records provide for detailed instructions and fill-in-the-blank opportunities for data entry, just like a Primary Manufacturing Record. Secondary Manufacturing Preparation Records are used for processing events that require evidence for step-by-step operations such as clean-in-place or steam-in-place operations. The preparation records can be designed as stand-alone documents or they can be used in association with SOPs that provide the directives.

These records do not usually contain the rigor of process control found in the Product Manufacturing Records, in that they usually lack Bill of Materials, preclearance, and close-out activities.

Logbooks

Logbooks are used throughout Manufacturing to provide for the chronology of events in a given processing area (see Chapter 6). There are logbooks for:

- clean rooms or product processing areas, providing a history of cleaning, testing, usage, maintenance, etc.;
- autoclaves, lyophilizers, hot air ovens, stopper washers, etc., detailing the batches or cycles of components that have been processed through those units;
- preparation of disinfectants used for cleaning production processing areas; and
- staging areas, detailing the movement of materials in/out of the area.

The Documentation Process in Production

Original, raw data is created when someone writes down something that is personally seen, heard, or detected. Original, raw data is the result of an original observation by a person or a sensor in an equipment system that records that data. Original, raw data is **not** an expectation of the results; it is **not** what you think; it is **not** a conclusion. Original, raw data is data that cannot be easily derived or recalculated from other information. For example, an operator is instructed to adjust the pH of a formulation solution within a range of 5.0 to 5.5 units. The operator records this event in the product record as "pH adjusted." This is not raw data. Raw data is the actual value that the formulation solution was adjusted to, such as "5.2".

Every individual who observes and records raw data on forms, product manufacturing records, logbooks, and/or collects data from automated equipment printouts is responsible for the accuracy, authenticity, and completeness of his/her data. When data is collected and an individual signs or initials the product or processing record, that signature should mean that the data:

- accurately describes what was observed;
- is authentic, meaning that these observations were made by the individual signing the document; and
- meets all expectations of the event, meaning that there is no unfinished work that would impact these observations.

Review of raw data by a second individual who is knowledgeable about the work is expected for critical processing events. This individual does not necessarily watch the work as it is performed or personally observe the events but is trained to audit and edit the data collection document. Consider the following responsibilities for those who review raw data:

- Is the data recorded properly (i.e., in the right place and in the right format)?
- Are all identity numbers correctly written on the data collection document?
- Is the document signed/dated by the individual who performed the work or made the observations?
- Is the data legible, logical, complete?
- Does the document reflect company policy about cross-outs, signatures, significant figures, averaging and uncompleted fill-in-the-blanks?
- Has the technician indicated that something "met specifications" when it did not? Are calculations correct?

Verification signatures generally refer to a second individual who actually observed the work performed by the first individual. This type of signature is required for only certain activities, such the weighing of formulation ingredients in drug or biologic manufacturing.

Making Changes to Original Data "Afterwards"

All changes to original data should be signed and dated accurately. Do not pre- or post-date signatures. The original entry should always be legible or visible through the cross-out and if the date of correction is after the date of collection, a rationale should be provided near the change or attached to the record. If the individual making the correction is not the individual who collected the data, then it is suggested that the originator of the data also sign the correction. If the change is significant, meaning that it impacted a product-release decision, then this event should be a candidate for review by the Change Review Board or another similar group within the organization.

Making changes to a document after the data has been collected requires the original data collection document and not a copy. If copies are kept routinely, then new copies need to be generated or existing copies changed in accord with the original. Having copies on file that do not match the originals is not good business practice.

Printouts

Identify and sign all printouts or charts from equipment used in data collection. For multipage documents, a single signature is sufficient only if there is pagination and complete identity of the documents page to page. If printouts are made on heat-sensitive paper, determine a way to make official copies of these printouts to be labeled as original data.

Contract Manufacturing

When product manufacturing processes are conducted by a contractor, the sponsor of the product remains responsible for the quality of the work performed. It is recommended, therefore, that the Production Department be involved in establishing the commitments of the contract and in the auditing of the contractor. The quality of both operations (the contractor and the sponsor) should be equivalent.

Documents and Records of Production

Documents that support the routine operations of Production must be available at or near the place where they are used. Given the problems with paper in most manufacturing environments, procedures and work instructions are more commonly appearing on monitors. Whatever the method of delivery, the most current version of each document should be the only one available for use.

SOPs written to support production are specific to the processing required. SOPs usually address the work instruction required to operate, clean, assemble, set up, and disassemble equipment.

The records of production events are usually kept by Quality Assurance. Product Manufacturing Records and completed logbooks, for example, are archived in a controlled access area under the control of QA. If Production needs files for reference, these are usually copies.

Tofte Medical, Inc.

Page 1 of 10

DXH-PLXC-0876; Rev 1
Labeling Batch Record

Lot # _____

Labeling Batch Record: Miracle Cure

Labeling Batch Record Release/Approval:

_____ QA _____ Date _____ Lot#

38–42,000 units = Batch Size

Bill of Material Record: Labeling Operations
Miracle Cure, 2ml Syringe

Product ID # SIP0-0100

Item	Item Spec. Code	Item Lot # Preparation Record ID #	Qty. Required	Qty. Issued from Inventory	By (MH)	Qty. into Production Area (Prod.)**	Recorded by (Prod.)	Verified by (Prod.)
Unlabeled Syringe of Miracle Cure	IPO007		40,000			**AA ___		
Tray Label	IPO222		40,000					
Syringe Label	IPO234			Roll #1 = ___				
				Roll #2 = ___				
				Roll #3 = ___				
				Roll #4 = ___				
				Roll #5 = ___				
			40-42,000 Total	Total = ___		**BB ___		

Exhibit H

Tofte Medical, Inc.

DXH-PLXC-0876; Rev 1
Labeling Batch Record

Lot # _____

○ **Preparation and Processing Clearance**
(1.0 - 4.0 can be prepared in any order or at the same time)

1.0 Area Set-Up

❑ Ensure that the area is clean and clear of materials from other labeling operations.

❑ Identify the area with the product ID # and Lot #.

❑ Identify and locate a **label reject form and/or bin** (labeled as "**OO**") on the line.

❑ Identify and locate a **syringe reject bin** (labeled as "**LL**") on the line for any units rejected *before* labeling operations

Performed by		Date	

2.0 Equipment Set-Up

○ ❑ Confirm that the Hot Stamp unit, Equipment ID # 23-5678 has been setup according to SOP MT8921, and that set-up is acceptable and documented in the Labeling Area Logbook.

Verified by		Date	

3.0 Labeler Set-Up

❑ Confirm that the Newcastle Labeler, Equipment ID # 23-9999 has been setup with the labels according to SOP MT0098. Confirm the following setting:

Observed Belt Speed Setting	_____ units	Performed by		Belt Speed Specifications = 30–50 units/minute
Observed Counter Setting	_____	Performed by		Counter Setting Specification = 0
Date of Fill	_____	Performed by		Counter Setting Specification = 0
2 years from Date of Fill	_____	Performed by		This equals assigned Expiration Date

Verified by		Date	

Exhibit H

Tofte Medical, Inc.

Page 3 of 10

DXH-PLXC-0876; Rev 1
Labeling Batch Record

Lot # _____

❑ Place an example label below

Performed by		Date	

4.0 Material Staging

❑ Move materials listed in Bill of Materials (BOM) into Labeling Area.

❑ Record receipt of materials on BOM by signing and verifying the entries, ensuring that the identification and number of units delivered matches.

Performed by		Date	

5.0 Clearance of Processing Area

Processing Area = K34

Preclear the area by ensuring that:

❑ the area has been cleaned according to SOP QA3456;

❑ the area is clear of other lots of similar products;

❑ the area logbook entry for cleaning is complete;

❑ the area is properly labeled for the event;

❑ all of the materials designated on the BOM are in the area;

❑ the actual lot number printed on naked syringes matches BOM;

❑ the preparation event confirmations, listed above, are complete;

❑ environmental conditions in the labeling area are acceptable;

❑ operators in area are properly gowned and trained for the duties they are performing; and

❑ this labeling event is entered into the area log.

Initial Processing Clearance Performed by (Supervisor)		Date		Time	

Exhibit H

Tofte Medical, Inc.

DXH-PLXC-0876; Rev 1
Labeling Batch Record

Lot # _____

Labeling Process

6.0 Begin Labeling

6.1 Follow the procedures for operation of the labeler in K34. As labeling proceeds, do the following as needed:

- Record *all* events in the Labeling Process Log.
- Label all completed trays with in-process labels. Attach label below.
- Place any syringes that are damaged and rejected from the labeler *before* labeling operations in a "**Syringe Reject Bin**" = **LL**
- Place all rejected labels that are free of the syringes on a waste label control sheet BB-VV-015 and/or in a reject bin = **OO**
- Remove QA samples from the line, *after labeling* and **always record # of samples taken from the line that are not returned**.

Labeling Start Date = CC		Time	

Performed by	

Tofte Medical, Inc.

Page 5 of 10

DXH-PLXC-0876; Rev 1
Labeling Batch Record

Lot # _____

Labeling Process Log

Date	Time	Start (S) Stop (X) Note (N)	Reason S = SMP R = roll change B = break P = new Preclearance M = mechanical E = end of shift T = test and setup X = see comm O = operator change	Comments	Recorded by						
		S	Initial start-up	Operators on duty =							

Exhibit H

Tofte Medical, Inc.

DXH-PLXC-0876; Rev 1
Labeling Batch Record

Lot # _____

6.2 Record the following:

Syringe Label Counter Value at End of Initial Labeling Event	_____ = KK	Observed by		Verified by	

Performed by		Date		Time	

7.0 End of Labeling

7.1 Check the line to ensure that all product and labels are removed from the equipment and area and that they are all segregated into their respective locations for: rejected labels
rejected syringes
acceptable, labeled syringes

Performed by		Date		Time	

7.2 Perform the following:

❑ Move labeled syringes available for packaging to cold room (2-8C).
❑ Ensure that they are labeled with Lot # and a Quarantine Label.
❑ Make a logbook entry for end of labeling operation.
❑ Identify area and equipment as available for cleaning.

Performed by		Date		Time	

7.3 Record Labeling End Date/Time

Performed by		Date of Labeling End = DD		Time	

Exhibit H

Tofte Medical, Inc.

DXH-PLXC-0876; Rev 1
Labeling Batch Record

Page 7 of 10

Lot # _____

Labeling Close-Out

8.0 Perform Syringe Reconciliation

# of Syringes Transferred to MH = KK		Recorded by		Verified by	
# of Rejected Syringes in Bin = LL		Counted by		Verified by	
# of QA Samples NOT Returned to the Line = MM		Counted by		Verified by	
Total Syringes Accounted for = KK + LL + MM =	_____ = NN	Calc. by		Verified	

# of Syringes/Syringes Brought into are on BOM **AA	_____ = AA	Recorded by		Verified by	
AA/NN x 100 = % Reconciled	_____	Calc. by		Verified by	

Specification for Product
Reconciliation = 99.9–100.0%

(   ) Syringe Reconciliation meets Specification; proceed.
(   ) Syringe Reconciliation does NOT meet Specification;
      write a Deviation Report and proceed.

Performed by		Date	
Verified by		Date	

Exhibit H

Tofte Medical, Inc.

DXH-PLXC-0876; Rev 1
Labeling Batch Record

Lot # _____

9.0 Label Reconciliation

KK # of Labeled Syringes Transferred to Packaging		Recorded by		Verified by	
OO Total # of Rejected Labels		Counted by		Verified by	
MM QA Sample Labels Removed from Line and on this PBR		Counted by		Verified by	
PP QA Sample Labels Syringes or on this PBR		Counted by		Verified by	
KK + OO + MM + PP = Total # of Labels Used	_____ = RR	Counted by		Verified by	

# of Labels Brought into are on BOM **BB	_____ = BB	Recorded by		Verified by	
RR/BB x 100 = % Reconciled	_____	Calc. by		Verified by	

> **Specification for Label
> Reconciliation = 99.9–100.0%**

( ) Label Reconciliation meets Specification; proceed.
( ) Label Reconciliation does NOT meet Specification;
     write a xxx Deviation Report and proceed.

Performed by		Date	
Verified by		Date	

Exhibit H

Tofte Medical, Inc.

Page 9 of 10

DXH-PLXC-0876; Rev 1
Labeling Batch Record

Lot # _____

10.0 Determine the Number of Labels to Return to Inventory

# of Labels Brought into are on BOM **BB		Recorded by		Verified by	
# of Labels Used During Labeling Operation = RR		Recorded by		Verified by	
# of Labels Returned = BB – RR =		Calc. by by		Verified by	
Labels Moved to Material Handling		Performed by		Verified by (MH)	

11.0 Calculate Product Yield

Yield = # syringes released for packaging operations (**KK**)
divided by
# of naked syringes brought into area (**AA** on Bill of Materials)
× 100

Record ** AA for syringes/syringes		
Record KK		
Actual Yield = KK/AA x 100 =	_____/_____ x 100 =	_____ %
Expected Yield Specification from = 99.9-100.0%		

(   ) Actual yield meets Expected Yield Specification; proceed.
(   ) Actual yield does NOT meet Expected Yield Specification;
write a Deviation Report and proceed.

Performed by		Date	
Verified by		Date	

Exhibit H

Tofte Medical, Inc.

Page 10 of 10

DXH-PLXC-0876; Rev 1
Labeling Batch Record

Lot # _____

12.0 Determine Processing Time

Record Labeling Start Time (see CC)	Date	Time
Record Labeling End Time (see DD)	Date	Time
Determine Total # of Hours for Processing End Time–Start Time	_____ Hours	
Time Specification = 8-16 Hours	_____ Hours	

( ) Actual Processing Time meets Processing Time Limits; proceed.
( ) Actual Processing Time does NOT meet Processing Time Limits; write a Deviation Report and proceed.

Performed by		Date	
Verified by		Date	

13.0 Summary and Batch Acceptance Criteria for Labeling

Labeling Lot # _____	Released _____	Units for Further Processing
Date of Transfer _____	Yield = _____	%

❏ This section of the batch record is complete with attachments.
❏ All processing equipment has been labeled as available for cleaning.
❏ Deliver this record to area Supervisor with attachments.
❏ Any deviations have been reported to Deviation Management.

ID # for deviations are: _____
( ) All acceptance criteria have been met.
( ) Deviations have been noted and supervisor notified.

Performed by		Date	

Exhibit H

13

Validation Documentation Basics

Aᴌᴛʜᴏᴜɢʜ ᴛʜᴇ ᴅᴏᴄᴜᴍᴇɴᴛs ᴀɴᴅ ʀᴇᴄᴏʀᴅs that support validation should be managed just like any other controlled document, it is not unusual to find that they are "somehow different." This distinction comes from the frenzy of facility start-up, and/or the confusion of thinking that validation is a one-time event. The issues that make the protocols and reports difficult to write/approve are derived from issues associated with the Validation Department's roles and responsibilities within the corporation.

When the Validation Department is small, which is usually the case in small, start-up companies, the individuals responsible for designing and implementing validation protocols have no authority over the individuals expected to do the work of validation (Maintenance, QC, Production, Material Handling). This organizational issue is evident in the documents, when writing a validation protocol becomes a multidisciplinary nightmare of review and approvals. If and when contractors are used to perform validation activities, this confusion of authority and responsibility increases in complexity. Although it might seem easier to suc-cumb to these struggles for control and "just write" protocols and reports *outside* of the exist-ing document control system—don't.

"Validation study plans, validation study protocols, and validation study reports are the working documents of validation. When these documents are written and organized in a man-ner that facilitates a logical flow of work and when these documents establish scientifically sound studies, validation becomes a straight-forward task: easily planned, easily performed and easily documented."[3]

Validation Study Master Plans

A Validation Study Master Plan organizes the validation commitments of the corporation by defining the corporate commitment to validation, providing a common approach to study format and content and assigning responsibility. A former FDA investigator explains:

"Planning documentation is a reliable predictor of GMP problems. During the initial phase of an FDA audit, it is customary to request the firm's validation plan documents. Management reaction to such a request often predicts the quality of the firm's documentation. If the firm does not have a formal written validation plan, then it is impossible for the system to be in a state of validation. Likewise, systems are suspect if the validation plan is poorly written, incomplete or unorganized, or if it lacks objectives or criteria. Conversely, a well-written plan reflects favorably on the overall quality of a firm's program."

- GMP Documentation Requirements for Automated Systems:
Part II, R.F. Tetzlaff, Pharmaceutical Technology, April, 1992.

"Planning documents also support good business practice. In the absence of good planning, work directed in procedures, protocols and programs can be redundant, inappropriate and inefficient. Planning documents should facilitate the consistent decision making that is fundamental for controlling medical product development, manufacturing and testing commitments."[3]

Validation Master Plans should include:

- an overview of the entire validation operation, the philosophy of validation, the approach, the regulatory requirements and the personnel structure of the validation organization;
- any definitions of approach for retrospective, concurrent or prospective validation;
- re-validation requirements;
- training requirements for individuals designing validation protocols, writing reports or performing validation activities;
- any rationale used to determine when validation and qualification are appropriate or required;
- any rationale associated with a matrix approach to validation of product types, processes, equipment, etc.;
- lists of actual systems and processes requiring validation;
- key acceptance criteria that applies to all validation studies;
- validation study protocol format, content and review/approval requirements;
- validation study report requirements that will apply to all reports;
- planning and scheduling responsibilities; and
- a list of associated documents that support the validation effort.

Validation Practices and Procedures

Qualification of Equipment

The terminology of equipment qualification includes installation qualification and operational qualification. Every company approaches equipment qualification with a different set of evaluation tools. What remains constant, however, is that qualification of equipment and associated controlling software occurs before the equipment is used to support an associated process validation. The level of detail associated with equipment qualification depends on the usage of the equipment. Equipment qualification studies cannot be adequately designed, therefore, without knowledge of the equipment's intended use. The basic principles of equipment qualification include:

- the equipment is installed in accordance with a plan, vendor requirements, local building codes, etc.;
- requirements for calibration, maintenance and cleaning have been developed as SOPs;
- operating requirements are established and tests conducted to assure equipment is operating correctly, under normal conditions (i.e., a set of conditions encompassing upper and lower processing limits that do not induce product or process failure); and
- operator training requirements are established and complete.

Installation Qualification Studies

The purpose of this evaluation is to confirm that the equipment is adequately designed and installed for its intended use. This requires knowledge of the user requirements for the equipment, meaning that one needs to link equipment performance to its intended use in routine processing. Consult Exhibit I.

The requirements for equipment installation qualification (IQ) must be determined or designated before the equipment is installed. These requirements usually include an installation plan or process. Once completed, the installation qualification is performed to confirm that the equipment, as installed, meets the expectations of the plan and the associated requirements. Consult Exhibit J.

After the initial installation event, the IQ record serves as an equipment specification, providing a convenient location for a description of critical features of the equipment (i.e., any component or feature which, if changed, could seriously affect the operation, performance or safety of the equipment or the product).

Consider the following information when designing an IQ record:

- "as built" drawings of the equipment or system
- materials of construction
- process flow diagrams for equipment
- equipment identity characteristics, including the presence of all associated equipment features
- utility requirements: feed and exhaust utility quality requirements as well as piping composition and diameter

- calibration of measuring devices that cannot be performed during OQ
- identification of cleaning, passivation, regeneration or any other procedures that must be performed before the equipment is operated
- safety features
- reference to operator or maintenance manuals
- reference to vendor support services or parts suppliers

IQ records should be completed by either a member of the Maintenance staff or the individual who will be responsible for the maintenance and performance of the equipment. The IQ requirements should be designed by someone different from the individual who performs the IQ assessment. IQ should be complete and approved before Operational Qualification begins. See Exhibit G.

Operational Qualification Studies

Operational qualification (OQ) study protocols confirm that the equipment operates as expected under ideal conditions; the protocol should define study conditions that prove the reliability of critical operating variables. Operational qualification study protocols must include what will be done, how it will be done, acceptance criteria and information on how raw data is handled or processed. Activities usually associated with operational qualification studies include:

- calibration of sensors or measuring devices
- qualification of associated
  - support processing such as equipment cleaning, disinfection, passivation, or sterilization
  - monitoring or controlling software
  - test methods used to assess performance
- systematic demonstration of equipment electromechanical features and functions
- demonstration of cycle performance, when appropriate, including the performance of programmed logic controllers or software programs
- demonstration of process uniformity or consistency
- demonstration of safety features and reset procedures
- demonstration of consistent performance can also be provided, if consistency will not be demonstrated during any associated process validation events
- completion of SOPs for routine equipment operation, maintenance, cleaning and calibration

Process Validation Study Protocols

When equipment qualification and process development is complete,[3] then the processes that the equipment supports can be validated. Process validation studies demonstrate that the established process can be performed reliably in the environment of intended use by its intended users and meet its processing specifications and the requirements of the process user. Good process development is no substitute for process validation studies and process validation studies are no substitute for a lack of process development data. Process development studies **determine** process control requirements; process validation studies **demonstrate** that process control requirements can be met reliably in the environment of use.

Process validation studies require that the processes be developed and established in controlled documents before the validation study begins. Documents that establish processes include Master Batch Records, Device History Records, and Test Method Procedures.

Protocols are not SOPs. Process validation protocols define a specific study that is used to determine the acceptability of a process; they are performed once or for a limited amount of time. There are many types of protocols used in development, manufacturing, and testing of products (e.g., development test protocols, stability study protocols, clinical study protocols). Process validation protocols are just one type.[1]

Process validation study protocols describe the study conditions required to demonstrate acceptable process performance. All process validation studies:

- should establish the study conditions, assure the continued adequacy and reliability of equipment, test methods, test systems, software, technicians, etc., *and* establish study acceptance criteria.
- must be reviewed and approved *before* the work begins.
- require a demonstration of process consistency in the challenge conditions of the environment of intended use. It is generally considered acceptable that three consecutive batches produced from processing that is within the established parameters of operation is considered sufficient for demonstrating consistency in validation events.
- require that the product of the event meets it established specifications.

See Exhibit K for an example of a process validation protocol and consider the following format and content guidelines for a process validation protocol.

1.0 Study Hypothesis/Study Question

Declare the purpose of the study as a hypothesis to be proven, or a question to be answered. For example,

> *Study Hypothesis: "The sterile filtration process established in MBR 57, version 04, can reliably produce a sterile formulation of Protein 33 that meets its associated specification PN2357, version 07."*

2.0 Study Objectives

Declare validation study objectives in measurable terms. Study objectives must be written in a manner that, when met, provide evidence that proves the study hypothesis. For example,

> *"Demonstrate that all process controls meet processing control testing limits, consistently, and that the formulation from the process meets its predetermined specifications for:*

- *three different lots of filters*
- *two different shifts of production*
- *two different filter manufacturers*
- *two different filtration assembly apparatus"*

3.0 Responsibility

Declare who is responsible for fulfilling the directives of this protocol. Define the intended user of the process and the intended user environment.

4.0 Definitions

Specifically define any terminology used in the study hypothesis and study objectives statements. Include any other definitions that will clarify or simplify the directives of the protocol.

5.0 Pre-Study Requirements

Describe any work that must be completed or reviewed before the study begins. Equipment and software supporting the process (such as filtration integrity testing equipment) may require calibration or qualification, technician training might need to be completed, and the SOP or manufacturing record which describes how to perform and control the process, must be approved.

6.0 Study Configuration and Conditions

Describe the study, i.e., specifically cite what will be tested, the methods that will be used to test, test or study observers and the handling of the test data to determine whether or not study objectives are met.

6.1 Sample Requirements and Sample Identification

What will be tested? How many samples are required or how much product/material is required to complete this study? How will samples or product be labeled and/or segregated during the study?

6.2 Test Parameters; Test Methods

How will the hypothesis be tested i.e., how will the objectives of the study be assessed? What product/material/sample parameters will be measured/observed? What are the test methods?

6.3 Test Acceptance Criteria

What is considered an acceptable result to meet study objectives?
Who is responsible for testing?

6.4 Testing Schedule

If a schedule is required to fulfill the directives of the protocol, what is the schedule and who is responsible for ensuring that the schedule is met?

6.5 Sampling Requirements

Describe the amount of product/material required for each sampling event, sampling techniques, sampling locations, and sample handling requirements.

6.6 Data Handling

When testing is completed for a sample point, how is the raw data managed? Are replicate values averaged or considered individually against acceptance criteria? Who is responsible for data calculations?

When testing is completed for the protocol, indicate how the data is managed and who is responsible.

7.0 Study Acceptance Criteria

Indicate what observations, values and results will be considered acceptable to prove the hypothesis or meet the requirements of the study hypothesis.

Records of Validation

Process Validation Study Reports

Validation study reports are generated after the work directed by a validation study plan or protocol is complete. Every validation study report must reference the specific validation study protocol identification number, and the revision level. These reports summarize data and state whether the acceptance criteria were met. When failures occur, these reports document the failure and provide a summary of the ensuing investigation. All data summaries should provide reference to the source of original data. Validation study reports must be signed and one signature must be responsible for the accuracy of the information presented in the report.

Process validation studies are used to "release" processes for use in commercial operations. As a result, processes are not considered validated until the reports are completed and approved.

Validation Study Notebooks/Validation Study Files

Validation study protocols, study reports, and raw data should be stored together in a format that facilitates their review by auditors. These validation packages must be secure. Validation study failures must be documented, reported, reviewed (investigated), and stored with the same rigor as study successes.

Procedures, Logbooks, Forms

The majority of documents used in validation are documents created for routine processing in other areas of the operation. The purpose of validation is to exercise these procedures, as defined in these documents, and demonstrate that they can be performed acceptably. There are a few lists, logbooks, etc., created to coordinate validation activities, but very few department-specific procedures.

Tofte Medical, Inc.

**Installation Qualification Records:
Creation, Use, Review & Approval**

IQ 23; 01

1.0 Purpose/Scope

This procedure describes the creation, distribution, use, review and approval of Installation Qualification Records (IQRs). This procedure applies to all IQRs created and used to support TMI.

2.0 Responsibility/Training

The creation, use, and approval of IQRs is the responsibility of the Validation Department. The authority to perform an IQ study can be handed down from Validation to qualified individuals (e.g., Maintenance and Engineering personnel or qualified outside contractors). All completed IQRs are reviewed and approved by Validation personnel before any further study using that equipment (e.g., operational qualification studies, performance qualification studies or validation studies) can be initiated. Individuals who contribute to the IQ studies that are not TMI employees must be trained in these procedures.

3.0 Procedures

3.1 Safety Precautions

When interacting with equipment, all personnel should adhere to the gowning and safety requirements for the area in which the equipment is installed and honor any vendor or supplier warnings about safety. Given that the IQR will verify some of the safety requirements for the equipment, it is important that individuals performing IQ studies be trained in the potential problems associated with the equipment before they interact with it.

3.2 Introduction/Preliminary Operations

The purpose of an Installation Qualification (IQ) study is to ensure that equipment is adequately designed and installed to meet its intended use. The IQ establishes the specifications for equipment and equipment systems that support critical processing. Equipment specifications are derived and determined by individuals knowledgeable in the use, maintenance, and calibration of the equipment. Equipment specifications are established in an IQR.

When established, the IQ study (i.e., the testing, inspection and verification) is performed by another individual. Results of this evaluation are recorded or referenced in the IQR.

Exhibit I

Tofte Medical, Inc.

**Installation Qualification Records:
Creation, Use, Review & Approval**

IQ 23; 01

3.3 Flow of Work

3.3.1 Initiating an Installation Qualification Study of Equipment or Equipment Systems

When the Validation Department determines that a new equipment installation requires qualification or change to an existing equipment installation requires requalification, a study is initiated by requesting appropriate records and forms from Document Control.

> *Note: Determining the need for the IQ is the responsibility of Validation Department; the authority to request IQ forms and records and to initiate IQ studies can be delegated to Maintenance/Engineering personnel and contractors, when appropriate.*

If the equipment to be qualified is a complex equipment system, Validation should complete a Equipment System Definition Form (Form #27). This form will list the individual IQRs required to study the entire system and will facilitate system review and disposition by Validation and QA. This form stays in Validation while IQ studies are conducted.

If the equipment to be qualified is not a complex system or when the individual IQ studies have been determined from a System Definition Form, request Installation Qualification Records from Document Control. To obtain a record, the requester must provide the name of the equipment to be qualified and the equipment ID # to Document Control.

The IQR form is produced by Document Control on blue paper. Each record is assigned a IQR # which is written on the document by Document Control personnel. Document Control maintains a log IQR issue events, recording the equipment to be qualified, the equipment ID #, the assigned IQR #, the date of issue and who requested the document.

IQR #s are sequential numbers (e.g., 97A001) based when the IQR was issued from Document Control. In this format, 97 is the year, A is for January, and 001 is the first IQR issued in January 1997.

Exhibit I

Tofte Medical, Inc.   Page 3 of 4

Installation Qualification Records:
Creation, Use, Review & Approval   IQ 23; 01

3.3.2 Completing the Installation Qualification Requirements and Specifications Section of the IQR

The "specifications" section of the IQR must be completed by
an individual:

- with knowledge about the specific purpose of the equipment
  in the TMI facility and/or with TMI processes;
- familiar with TMI and current industry practices for installation,
  maintenance, and calibration of such equipment; and
- with access to manufacturer and vendor information about the
  equipment and equipment system.

Requirements and specifications should consist of equipment
attributes, features, and characteristics that—*if they were not
specified and met*—the equipment would not perform safely or
would not perform adequately for its intended use. Whenever
possible express specifications as a range of operational values.
Points to consider when establishing requirements and specifications
are provided in Appendix A of this procedure. The individual who
establishes these requirements is responsible for the completion of
pages 1and 2 of the IQR form. This individual can delegate and
initiate the work of checking the accuracy of documentation before
the IQ study proceeds or the R/S section is completed.

Complete the left-hand side of pages 3 to 5. These handwritten
entries should be legible and clearly stated.

3.3.3 Review and Approval of the IQR Requirements and Specifications

When the Specifications section of the record has been completed,
this portion of the record is reviewed and approved by another
knowledgeable individual. The review and approval of the
Specification section must be performed by a TMI employee.

Exhibit I

**Installation Qualification Records:
Creation, Use, Review & Approval**

IQ 23; 01

3.3.4 Performing the IQ Study

When the Specification section is approved, the IQ study is performed. The IQR form is given to an individual who verifies if the requirements and specifications and/or performs calibration, cleaning, passivation work. When the work directed by the IQR requirements and specifications is completed, the results are indicated or referenced in the "as found" section of the IQR (the right-hand side of the record, pages 3 to 6). The individual performing the study should not be the same individual who determined the requirements and specifications.

3.3.5 Review and Disposition of the IQ Record

When the "as found" section of the IQR is completed, that individual signs page 6 of the IQR and delivers the document to the Validation Department for final review and approval.

A disposition of "complete and acceptable with corrections noted" can be used *only* if the corrections initiated will not directly affect the performance or configuration of the equipment. For example, a CAD drawing scheduled for update or a drawing correction that has been scheduled but not completed would not affect equipment performance or configuration. This disposition option facilitates further study of the equipment or the processing it supports (i.e., Operational Qualification or Process Validation). The work initiated, however, *must* be completed before any associated Validation report is considered acceptable.

The Validation Department and QA are responsible for verifying that the "as found" conditions fulfill the established requirements and specifications, and determining the disposition of the IQ study.

3.3.6 Filing and Distribution of IQRs

The Validation Department retains the original paperwork associated with the IQR. These records are archived in association with the validation study reports that they support.

Exhibit I

Tofte Medical, Inc.

Page 1 of 6

Installation Qualification Records

IQR # _____

Installation Qualification Record

Equipment ID # _____

Equipment or equipment system name: _____

Equipment location _____, **or ( ) portable.**

Is this equipment part of a larger equipment system?   ( ) No   ( ) Yes

Equipment operated by which divisions/departments:
( ) Bulk Production
( ) Aseptic fill
( ) Material Mgmt.
( ) Packaging
( ) Development
( )
( ) Maintenance
( ) QC

Functional Requirements for Equipment: *(Describe purpose of equipment, output requirements, processing that it is designed to or is likely to support.)*

_____

_____

_____

Basis of Operation, Operational Features, Operational Limits: *(Provide a general overview for the basis of operation of the equipment and a flow of operations; attach diagrams when appropriate.)*

_____

_____

_____

Form #76, Rev. 0

Exhibit J

Tofte Medical, Inc.

Installation Qualification Records

IQR # _____

Equipment Identity

Purchase Order # _____

Purchasing Specification # _____

Manufacturer: _____

Supplier/Vendor: _____

Model # _____

Serial # _____

Asset # _____

Date of Installation: _____

Requirements

As Found Observations

Documentation Requirements (prints, manuals, drawings)	Accuracy Check required? (yes/no)	Document ID	# Copies	Acceptable Accuracy? (yes, yes with corrections)	Sign/Date

*Actions initiated to correct document deficiencies _____

Form #76, Rev. 0

Exhibit J

Tofte Medical, Inc.　　　　　　　　　　　　　　　　Page 3 of 6

Installation Qualification Records　　　　　　　　　IQR # _____

Equipment Specifications

Specifications	As Found Observations									
Production Capabilities or Capacity Specifications:										
Safety Features:										

Form #76, Rev. 0

Exhibit J

Tofte Medical, Inc.

Page 4 of 6

Installation Qualification Records

IQR # _____

Equipment Specifications

Specifications	As Found Observations								
Utility Specifications:									
Product Contact Surface Requirements									
Spare Parts and Interchangeable Parts Required									

Form #76, Rev. 0

Exhibit J

Tofte Medical, Inc.

Page 5 of 6

Installation Qualification Records

IQR # _____

Off-Line Requirements

Calibration
(list item and ID #)

Date of Calibration	Calibration Acceptable?		Signature/Date
	Yes	No	

Cleaning and Passivation
(list item and method used)

Date of Work	Records Reviewed and Acceptable?		Signature/Date
	Yes	No	

Critical Material Inspection and Testing Requirements
(list requirement and method used)

Date of Work	Records Reviewed and Acceptable?		Signature/Date
	Yes	No	

Form #76, Rev. 0

Exhibit J

Tofte Medical, Inc.

Page 6 of 6

Installation Qualification Records

IQR # _____

Equipment Component Data Sheets that support this IQR

Component _____ Equipment ID # _____

Component _____ Equipment ID # _____

Component _____ Equipment ID # _____

Component _____ Equipment ID # _____

The Requirements and Specifications Section of the Installation Qualification Record

Completed by _____ Date _____ Affiliation _____

Reviewed by _____ Date _____

The Installation Qualification "Observations" and "As Found" Information

Completed by _____ Date _____ Affiliation _____

Disposition of Installation Qualification Record

( ) Complete and acceptable.

( ) Complete and found unacceptable; make required changes and repeat IQ study.

( ) Complete and acceptable with corrections noted.

Note: These corrections do not directly affect the installation or performance of the equipment; associated studies can proceed but corrections must be complete before any associated validation study reports based on this IQ can be found acceptable.

Corrections required _____

Validation Department _____ Date _____

Form #76, Rev. 0

Exhibit J

Tofte Medical, Inc.

Page 1 of 5

Validation Protocol VP43; 01

Sterilization of 20mm Stoppers in Superior Autoclave

1.0 Study Hypothesis

20mm rubber stoppers can be reliably and effectively sterilized for use in aseptic processing when sterilized in the Superior Autoclave as directed by cycle #G13 (SOP 222).

2.0 Study Objectives

- Demonstrate that the stoppers are exposed to the processing conditions, uniformly throughout the chamber.
- Demonstrate that the sterilization cycle, G13 in SOP 211, can reliably meet processing control parameters when programmed in the Superior Autoclave EQ 23.
- Demonstrate that the stoppers are of acceptable quality, (e.g., not sticky or discolored or otherwise adversely altered by processing).
- Demonstrate that this cycle can achieve a sterility assurance level of 10^{-6} with a full load of stoppers (PN 2134).

3.0 Responsibility

Validation is responsible for ensuring that this protocol is followed; Production and QC are responsible for performing the work.

4.0 Definitions

Full load of stoppers = 9 trays placed 3 to a shelf with each tray containing 15 to 18kg of 20mm stoppers

Sterility Assurance Level = the probability of a nonsterile unit

5.0 Pre-Study Requirements

5.1 Installation Qualification

Perform the Installation Qualification of the Superior Autoclave EQ23, as directed in SOP 021. Ensure that the IQR is reviewed by Maintenance and signed off as acceptable before proceeding.

Exhibit K

Tofte Medical, Inc.   Page 2 of 5

Validation Protocol VP43; 01

5.2 Operational Qualification/Preliminary Operations
- Calibrate thermocouples and RTDs according to SOP 222.
  Must meet criteria of +/- 0.2 C.
- Calibrate timing devices according to SOP 333.
  Must meet criteria of +/- 30 seconds.
- Calibrate pressure gauges/recording devices according to SOP 444.
  Must meet criteria of +/- 0.5 psig.
- Calibrate temperature recording device according to SOP 555.
  Must meet criteria of +/- 1C.

6.0 Study Configuration and Conditions

Perform two heat distribution studies on an empty chamber and one cycle of a full load of stoppers to demonstrate uniformity of heating and to determine the cold spot in the chamber. When the cold spot is determined, the controlling RTD and one thermocouple will be located in this position for all future cycles.

To demonstrate that a sterility assurance level of 10^{-6} can be achieved consistently, three consecutive cycles, with full loads of stoppers and three consecutive cycles of one tray of stoppers, all spiked with 10^6 spores of *Bacillus stearothemophilus*, will be run.

To demonstrate reliable equipment performance, it is expected that all processing control parameters will be met for all these demonstration cycles.

6.1 Heat Distribution/Empty Chamber

6.1.1 Placement of TCs and RTDs.

Place thermocouples in an X pattern throughout the chamber for each shelf and one in the drain. Controlling RTD is in the drain. Complete Form PR32 indicating the exact placement of the thermocouples and RTDs for each cycle.

6.1.2 Cycle Settings Cycle # G 13

Time setting = 30 minute
Temperature setting = 121.6C
Pressure = 14.7 psig
Prevac cycle = 3 minutes with 3 purges
Cooling cycle = 60 minutes

Exhibit K

Tofte Medical, Inc.

Validation Protocol VP43; 01

6.1.3 Cycle Data Collection Requirements

Operate the autoclave as directed in SOP 211. Enter the cycle in the auto-clave logbook; record the cycle number, date and operator initials on the cycle chart.

6.1.4 Heat Distribution Cycle Acceptance Criteria

- There are two consecutive heat distribution cycles.
- All thermocouples and RTDs are recording throughout at least two cycles; diagrams of TC and RTD placement are available for each cycle.
- All RTD and TC temperature recordings be 121.6 C +/- 0.5C for 30 minutes +/- 3 minutes; documentation is available to support these observations.
- Pressure recording devices must meet 14.7 psig +/- 0.5 psig for 30 minutes +/- 3 minutes; documentation is available to support these observations.
- Three consecutive cycles must meet the above criteria before proceeding.

6.1.5 Determination of Uniformity

All thermocouple readings should read within +/- 0.6C of the controlling RTD.

6.1.6 Identification of Cold Spots

Determine the cold spot by averaging each thermocouple reading throughout the 30-minute cycle time. If more than one TC has the lowest average temperature and one is in the drain, designate the drain as the cold spot for the final run. Otherwise pick the location where the TC is least likely to interfere with normal loading of the chamber.

6.2 Heat Distribution/Loaded Chamber

Repeat the cycle with a full load of stoppers. Ensure that the stoppers have been processed (i.e., washed and siliconized) as would occur routinely (SOP101).

The cycle must meet the same criteria as the empty chamber cycles. Confirm that the cold spot location remains the same, or change the location as the data indicates.

Exhibit K

Tofte Medical, Inc.   Page 4 of 5

Validation Protocol VP43; 01

6.3 Performance Qualification/Heat Penetration Studies

6.3.1 Preliminary Operations

- Ensure that the biological indicators (*Bacillus stearothermophilus* PN #1234) meet specifications (i.e., are validated).
- Ensure that the heat distribution studies are acceptable.
- Ensure that the cold spot, determined from the heat distribution studies, is monitored by a RTD.
- When using materials that will routinely be washed and/or processed before exposure to steam sterilization, ensure that these materials have been processed before they are used in these studies.

6.3.2 Heat Penetration Studies

Perform three consecutive runs of:

    Materials to be sterilized = 20mm stoppers, PN 2134

    Loading Configuration = full load

    Location of thermocouples = one in each tray near the spore strip;
        TC and RTDs in cold spot

    Cycle Setting = Cycle G 13

Perform three consecutive runs of:

    Materials to be sterilized = 20mm stoppers, PN 2134

    Loading Configuration = one tray located in center of center shelf

    Location of thermocouples = one in the stopper tray near the spore strip;
        TC and RTDs in cold spot

    Cycle Setting = Cycle G 13

Acceptance Criteria for all cycle runs:

- All available data from TCs and RTDs meet the following time/temperature requirements:

    121.6C +/- 0.5C

    30 minutes +/- 1 minute

    The raw data records are available to support these observations.

Exhibit K